Coping with an Idea of Ecological Grammar

PhilologicaWratislaviensia: From Grammar to Discourse
Edited by ZdzisławWąsik

Advisory Board
Jerzy Bańczerowski (Poznań)
Leszek Bednarczuk (Kraków)
Doina Cmeciu (Bacău)
John Deely (Houston)
Franciszek Grucza (Warszawa)
Richard L. Lanigan (Fairfax Station)
Norbert Morciniec (Wrocław)
Roland Posner (Berlin)
Daina Teters (Riga)

Volume 2

PETER LANG
Frankfurt am Main · Berlin · Bern · Bruxelles · New York · Oxford · Wien

Elżbieta Magdalena Wąsik

Coping with an Idea of Ecological Grammar

PETER LANG
Internationaler Verlag der Wissenschaften

Bibliographic Information published by the Deutsche Nationalbibliothek
The Deutsche Nationalbibliothek lists this publication in the Deutsche Nationalbibliografie; detailed bibliographic data is available in the internet at http://dnb.d-nb.de.

Published with financial support from
Philological School of Higher Education in Wrocław

Typesetting by Zdzisław Wąsik

Editorial proof by
Martin Dolan & Sylwia Rudzińska

Cover design by
Beata Opala

ISSN 1866-699X
ISBN 978-3-631-60228-7
© Peter Lang GmbH
Internationaler Verlag der Wissenschaften
Frankfurt am Main 2010
All rights reserved.

All parts of this publication are protected by copyright. Any utilisation outside the strict limits of the copyright law, without the permission of the publisher, is forbidden and liable to prosecution. This applies in particular to reproductions, translations, microfilming, and storage and processing in electronic retrieval systems.

www.peterlang.de

Contents

Introduction: From the ecology of language to ecological grammar of linguistic linkages 9

Chapter One: Language and man in the history of linguistic thought .. 15

1. Genetic and typological diversification of languages in compartive investigations 16
2. The autonomy principle in the study of language as an abstract system of communicative means 21
3. Homocentric and integrationist conceptions of the object of linguistic studies 24

Chapter Two: An inquiry into the functional view of language as a tool of communication or a property of task-realizing communities 29

1. The interdisciplinary background of functionalism as an investigative perspective 29
1.1. Functionalism searching for invariant features based on the principle of abstractive relevance 31
1.2. Functionalism exposing teleological or etiological explanations of change and adaptive processes 32
1.3. Functionalism referring to the dispositional properties of objects satisfying subjective needs of organisms and the systemic requirements of culture and society 33
2. Towards a paraphrasing explanation of the term *function* 36
3. A tool-oriented view of language functions realized by verbal means as one of the constituents of a communication scheme 38
3.1. Alan Gardiner's definition of function as a performance ability of language products to act in subservience to certain aims in the view of language users 39
3.2. Karl Bühler's representational function and the principle of abstractive relevance in the instrumental model of language 40

3.3. A critical review of Roman Jakobson's eclectic distinction of six functions of language as parallel to six communication constituents 42
4. A survey of functions attributed to language as a socio-psychological property of communicating individuals 45
4.1. Socio-pragmatic and cognitive perspectives on the inter-subjective functions of language in M. A. K. Halliday's and Robin P. Fawcett's depictions 45
4.2. Classifying linguistic functions as social domains of language use according to Charles Ferguson and William A. Stewart 48
5. Against the indiscriminate enumeration of linguistic functions 50

Chapter Three: The linguistic properties of communicating individuals and their role in the construction of an intersubjective world of meanings 55

1. Language as a creator of communicative collectivities 55
2. The linguistic properties of man against the background of constructivists' theories 56
3. On the personal constructs and social construction of reality 57
4. Man as the object of linguistic studies in Franciszek Grucza's and Victor H. Yngve's depictions 63
5. The accessibility of the linguistic properties of communicating individuals 68

Chapter Four: On the ecosemiotic existence mode of language in local and global linkage aggregations 73

1. On the notion of ecology and ecologism as an investigative attitude in the sciences of language 74
2. Specifying the concept of an ecological grammar of linguistic linkages realized in human communication 79
3. Presenting social groups in terms of ecologically determined dynamic systems 82
4. Linguistic ecosystems under the pressure of globalization and bioinvasion 86

Chapter Five: Sociological aspects of linguistic pragmatics in the light of hard-sciences 91

1. Physical and logical domains in the investigative field of human linguistics 91

2. How to do things with "sound waves" or with "words"? 92
3. Between philosophical and hard-science pragmatics 99

Conclusions: In favor of the notion of grammar as a network
of interpersonal and intersubjective linkages 105

Supplement One: Applying an ecological model of language
to the external characteristics of Frisian 115

1. The ecological status of Frisian in the Netherlands
and the Federal Republic of Germany 116
2. The state of the art in ecological studies on Frisian 120

Supplement Two: Exposing the markers of Frisian ethnicity
through a semiotic perspective 123

1. Myths and stories of the origin of Frisians as a reference source
of ethnic separateness 125
2. National consciousness, "pillarization" and secularization
of Frisian society 128
3. Language as a fundamental symbol of Frisian ethnicity 129
4. The ecological and societal-cultural symbols of Frisianness 132
5. Summarizing the semiotic markers of Frisian ethnicity 136

Supplement Three: Presenting West Frisians as an aggregation
of ecological linkages 137

1. Applying Victor H. Yngve's theory of human linguistics
to the analysis of interpersonal relationships among West Frisians 137
2. Observable and non-observable facts in human linguistics:
between hard and soft sciences 145

Bibliography: Works cited and consulted 147

Index: Names and terms from the main text 163

Introduction

From the ecology of language to ecological grammar of linguistic linkages

This book recapitulates the contents of my earlier presentations and subsequent publications associated with the topics which marked the way of my professional development in the domain of the sciences of language. I had the opportunity to acquaint myself with the issues of the broadly understood linguistics, considering the investigative aspects proposed by its neighboring disciplines, when I was engaged in the individual studies devoted to the methodology of the classification of European languages under the guidance of Professor Antoni Furdal at the University of Wrocław in the academic year 1983/84.

I was able to continue my interest in studying the properties of language from the perspective of its functioning in environmental conditionings when I was offered the position of Assistant in the Department of General Linguistics, as a result of competition which took place in the autumn of 1986. At that time it was suggested to me that I may undertake a theme connected with the ecological status of Frisian as a minority language. Having accepted this new investigative domain for my future doctoral dissertation, I started at first with methodological studies on the model of the holistic description of language both in the external and internal dimension, and I then tried to familiarize myself with the existing knowledge concernig the situation of Frisian as a minority language in Western Europe. To fulfill this task, I made contact with the Frisian Academy in Leeuwarden [Fryske Akademy, Ljouwert], and afterwards with the Center for North-Frisian Dictionary (Nordfriesische Wörterbuchstelle) in Kiel wherefrom I had received preliminary material data useful for the formulation of the main theses of my future dissertation. I had the opportunity to obtain the complete source materials for the empirical part of my work during my stay in the Frisian Institute at the State University of Groningen which was supported by a stipend of the Ministry of Education of the Kingdom of Netherlands (01. 02.–30. 06. 1990) where Geart van der Meer and Oebele Vries had been my scientific associates and collaborators.

After my return to Poland I prepared a dissertation on the ecology of minority languages (*Ekologia języków mniejszościowych na przykładzie języka fryzyjskiego*), which was defended at the Philological Faculty of the University of Wrocław on November 23, 1995. Antoni Furdal from the Department of General Linguistics assumed the duty of a professor conferring the doctoral degree. Norbert Morciniec from the Institute of the German Philology of the University of Wrocław and Leszek Bednarczuk from the Institute of Romance Philology of the Pedagogical Academy in Kraków acted as reviewers of this dissertation, on the basis of which I had received the title of Doctor of Humanistic Sciences on November 28, 1995. One of the exemplars of this dissertation in the form of a computer print had been sent to the library of the Frisian Academy in Leuwarden as a result of the interest expressed by Durk Gorter who was working in the field of the sociology of Frisian. On the basis of studies concentrated around the investigative domain of my dissertation, I was able to publish several separate articles devoted to the external characteristics of Frisian, classifications of the Germanic communicative communities, as well as from the domain of the methodology of general linguistics.

With the theoretical problems connected with the ecology of language I became familiar in Wrocław, Poland, before embarking upon a scientific trip to the Netherlands where I planned to collect the source materials for my dissertation. Ecological studies were developed as one of the current trends in the activity of the Department of General Linguistics at the University of Wrocław, starting from 1985. Its foundations constituted the linguistic theory of language as elaborated by Leon Zawadowski (1966) and the descriptive model of holistic description of language initiated in the second edition of the book on open linguistics by Antoni Furdal (1990 [1977]). The first announcement of this program, and the first investigative samples from the viewpoint of external linguistics based on the descriptive model of Einar Haugen, had found its place in the first volume of *Studia Linguistica* edited in the Department of General Linguistics under the editorships of Zdzisław Wąsik having been devoted to the issues on the ecology of language (1993). Within the framework of this model, I presented the results of my first explorations from the domain of Frisians and their language, in the form of a communiqué, at the meetings of the Scientific Society of Wrocław on January 26, 1988, pertaining to the situation of the Frisian language in the Netherlands and the Federal Republic of Germany (Lizis 1990). In addition to this, on December 14, 1988, I also delivered a lecture for the Society for the Ethnology (Towarzystwo Ludoznawcze) in Wrocław. Of crucial importance for the theoretical foundations of the future doctoral dissertation was my introductory article summarizing investigations on the external description of language, prepared for print in 1994, and published two years later (Lizis 1996). Elaborated in the doctoral dissertation against the background of hitherto existing typologies of ecological variables for the requirements of comparative linguistics, my own proposals had been presented at the international conferences in Poprad (1995), Gdańsk (1996), Berlin (1997), Poznań (1997) as well as at the First Congress of Modern Languages and Literatures in Łódź (I Kongres Neofilologiczny w Łodzi, 1998).

The final results of my studies including the ecological model for the external description of minority languages are available in the following publications on: (1) the domains and tasks of external description of languages: on the basis of Frisian (Wąsik, E. 1999a), (2) the ecology of Frisian: from the studies on the situation of ethnolinguistic minorities in Europe (Wąsik, E. 1999b), and (3) the contribution to the model for the ecological description of minority languages in Europe on the basis of Frisian (Wąsik, E. 1999c). They have been also extensively popularized in a monographic work on the Lemkos: within the scope of the research into ethnolinguistic minorities in Europe by Małgorzata Misiak (2006: 25–27).

The theoretical achievements of my doctoral dissertation contributed to the elaboration of the model of an external description of language encompassing certain sets of ecological variables. The work on the issues grouped around the ecology of language bore fruits in the neighboring topics, which found its expression in the presentation and related publications: (1) on the notion of eurolinguistics: a side note to the book of Norbert Reiter on the basics of Balkan studies (Wąsik, E. 1988), (2) on the metaphor of roofing in the context of the classification of Germanic languages (Wąsik, E. 1998), (3) the domains of language use from the perspective of external linguistics (Wąsik, E. 1999d), (4) on the necessity of teaching the native language in the conditions of diglossia: on the basis of Frisian (Wąsik, E. 2000b), (5) ethnic identity in a semiotic perspective (on the example of Frisian) (Wąsik, E. 2001), (6) the methodological aspects of minority-language studies (Wąsik, E. 2005). In this last position, the material sources have been taken into consideration not only from the domain of Frisian but also Kashubian of today.

The topic of my second dissertation, published in Polish with a summary in English as a separate book (Wąsik, E. 2007), in which I was trying to answer the question: Language – a tool or property of man? (Towards an idea of ecological grammar of human linkages), had been formed on the basis of my active participation in two thematic workshops, devoted to (1) "Ecosemiotics: Studies in Environmental Semiosis" within the frames of the *Nordic-Baltic Summer Institute for Semiotic and Structural Studies*, Imatra, 2000 as well as (2) "Exploring the Domain of Human-Centered Linguistics from a Hard-Science Perspective" at the *Societas Linguistica Europaea 33rd Annual Meeting*, Poznań 2000. As a result of this participation, some of my earlier beliefs (Lizis 1996) based on the theoretical assumptions elaborated in the Department of General Linguistics in Wrocław, which gave rise to the theoretical foundations of my doctoral dissertation on the ecology of Frisian published in 1999, had to undergo a thorough revision.

The journey to this new dissertation has been marked by my papers presented at various international conferences, and documented later on in articles and summaries in respective books of abstracts: "Towards redefining the concept of the ecology of language in the framework of human-centered linguistics (with special reference to Frisian-speaking linkages)" (Poznań 2000); "On the idea of an ecological grammar of verbal discourse from the perspective of human-linguistics" (Leuven 2001); "'Ecological grammar' of communicative linkages: Between bioinvasion and cultural transmission in the globalized world" (in co-authorship with Zdzisław Wąsik, Warszawa 2002); "On the idea of an 'ecological grammar' of verbal discourse from a human-centered perspective" (Poznań 2003); "Describing Frisian communities in terms of human linguistics" (London – New York 2004); "'Ecological grammar' of communicative linkages: Between bioinvasion and cultural transmission in the globalized world" (in co-authorship with Zdzisław Wąsik, Frankfurt am Main, etc. 2004); "Sociological pragmatics from a hard-science perspective. A sidenote to the conception of human linguistics" (Poznań 2005); "Linguistic properties of communicating individuals in the construction of intersubjective world of meanings" (Toruń 2006); "Applying Victor H. Yngve's theory of human linguistics to the analysis of interpersonal relationships among communication participants in social reality and literary fiction" (in co-authorship with Zdzisław Wąsik and Maciej Kielar, Poznań 2006); "On the notion of ecology and ecologism as a relativistic attitude in the domain of language sciences" (Bremen 2006).

What constitutes an apparent *novum* in the frame of reference of the second approach to the ecological view of human communication is connected with the placement of language among the properties of man. Its concrete existence mode should be found in observable interpersonal linkages created on the basis of how people communicate. The subject matter of the linguist's concern are, therefore, the inherent and relational properties of language that unite people as linguistic-communicational linkages with their natural and cultural environment assembling communication participants who send and receive verbal messages in the same way.

The study enters into the domain of epistemology and the historiography of language sciences in relation to the sciences of man. In such a human-centered perspective, any belief about the transparently ordered nature of language is rejected. The same refers to the possibility of its description in meaning-and-understanding-related terms. Language is not a homogeneous phenomenon. Its concrete existence modes are to be found in the behavioral signs of tasks being performed by humans. Observable are thus the properties of individuals communicating in a given language when they aggregate into communicative linkages sending and receiving verbal meaning bearers in the same manner. It is assumed that the boundaries of human linkages are fuzzy and haphazard depending on their ecological conditionings. The grammar of the linguistic communities formed on their basis has, therefore, occasionally changing centers and peripheries.

Chapter One

Language and man in the history of linguistic thought

The subject matter of our considerations constitutes those linguistic-communicational properties of man, which can be reconstructed from the concrete and assumed objects constituting the investigative domain of the linguistic and non-linguistic theories of language. The scope of our investigative material will be determined by the empiricist and rationalist perspectives on language, in which the properties of man are identified with its observable aspects in contra-distinction to inferred concepts, in which the linguistic nature of man is studied as the primary object of investigation. The study of the relationships between language and man will be based on historiographical accounts of the various practitioners and theorists of language studies. Moreover, we will pay special attention to those opinions advocating a human-centered perspective in language sciences and their neighboring disciplines.

In search of who the man is in relation to language, and what the language is in relation to man, one may notice that the linguistic properties of a speaking and understanding individual are presented either as his inherent qualities or as externally perceivable features of himself as an organism. Man is usually presented in linguistic works as an individual with concrete and mental traits, or as a representative of a communicative community, but always in relation to language. However, language is studied not only in relation to man but also in abstraction from man as a living being, a member of society and a conscious person.

The assumption that language (identified with types of sound waves produced in acts of speech) ought to be considered as a real object independent of people has to be confronted with the statement that what determines the nature of human communication constitutes exclusively the linguistic properties of a communicating individual. A critical review of a linguist's attitude towards the existence form of language is connected with the question of its approachability as a phenomenon underlying subjective cognition. Only linguistic facts which are observable in the physical domain can be approached with direct methods. Nevertheless, indirect cognition is impossible in the case of linguistic facts connected with interpersonal communication, the notification of which is exclusively based on the knowledge of observers, making inferences with respect to intentions, purposes or tasks of individuals.

The depiction of man with respect to his linguistic properties is dependent upon the view of whether the existence of language is accepted in terms of human capabilities realized in communicational acts or denied as a pure construct without any observable reality. Similarly, as to the definition of language, practitioners of linguistic sciences sometimes rely on a metaphorical presentation of man as a speaking animal, reducing in that way his image to a theoretical model.

Linguistics became an independent science after it had separated its own subject matter from the other human-related sciences as late as in the beginning of the 19th century. In the history of linguistic disciplines, there appeared, coexisted and disappeared, in succession and concurrence, three investigative layers: comparativism, structuralism and poststructuralism, thereby shaping its epistemological foundations by introducing various concepts regarding the nature of language and the ways how to approach it. In principle, they have oscillated between "logocentrism", which treats language as an object situated beyond the communicating subject, and "anthropocentrism" where man as *homo loquens* became an object of scientific studies standing in relation to language as one of his environmental conditionings and/or his individual subjective property.

1. Genetic and typological diversification of languages in comparative investigations

In the first three decades, the representatives of historical comparative studies, Friedrich (Karl Wilhelm) Schlegel, Franz Bopp, Jakob Grimm and others, discussed, *inter alia,* by Adam Heinz in his outline of the history of linguistics (1978: 133–140), as well as Robert Henry Robins in *A Short History of Linguistics* (1974 [1967]: 170–172), were interested in comparing languages by applying an inductive method. For this reason, generalizations about languages, as cognate or related, were made on the basis of correspondences between them. This attitude changed from the conviction that a certain group of languages must have a common ancestor to the postulate that primary languages can be reconstructed. The inductive method was replaced, in the next instance, by a deductive method. As stated by Heinz (1978: 168–169), generalizations made on the basis of evidence from Romance languages having Latin as their immediate primary language, allowed, for example, August Schleicher in the 60s of the 19th century to submit a hypothesis that the same fact may be inferred from the correspondences among Slavic and other Indo-European languages.

Initially, within the framework of comparativism, linguistic investigations were based on a philological method, and later on a historical-comparative method. Abstracting from man, researchers dealt with concrete textual data preserved in the written documents of dead languages. The purpose of such philological studies, apart from the emendation, explication and edition of texts, as sources of cultural knowledge, was seen in the reconstruction of the earlier phases of vanished languages. In the formation of a morphological object of linguistic studies and comparisons, imitating the distinctions and methods of botany and zoology, attention was paid to letters as representations of speech sounds, roots and stems of words against the background of their derivational and inflecting forms. Their meaning was deduced exclusively through the vantage point of hitherto known languages, in such cases where it was deemed useful for the translation of studied and edited texts.

In particular, historical comparative investigations, as summarized, *inter alia,* by Neville Edgar Oscar Collinge in his articles "History of comparative linguistics" (1995a: 198–200) and "History of historical linguistics" (1995b: 204–206) and Adam Heinz (1978: 128), aimed at confronting text elements of languages that belonged to different historical periods as well as different territories, and then attempted to draw conclusions with respect to the occurrence of linguistic phenomena in the subsequent developmental phases of a given language is concerned. The result was a classificatory approach to the history of languages. Moreover, the aim of such investigations was to search for the origins of languages and to explain the causes of language evolution as well as the reasons for language changes. It was noticed that languages do not only separate in various directions, but also that they unify and become similar due to contacts between the individual language bearers. In this context, the perspective of divergent evolutionism which was advocated by August Schleicher through the theory of a genealogical tree of language families had to be confronted with that of convergent diffusionism as proposed by his pupil Johannes Schmidt in the ethnological wave theory of language changes, as pointed out by Zdzisław Wąsik (1999: 59, see footnote 6) in his paper on language autonomy, the variability of languages and multilingualism from the perspective of comparative linguistics.

The investigative practice of the first comparatists was based on the principle that language bearers originally formed a homogeneous community, which then differentiated over time. Following the views of romanticists, evolution meant decadence, and in this particular case, the corruption of a language, which was originally characterized by simplicity and perfection. In accordance with the views of rationalists, however, because of an increasingly optimizing characteristic of the mind of man evolution was understood as progress. Basing their claims on linguistic facts, in abstraction from man, practitioners of comparative linguistics who tended to reconstruct a parent language of Indo-Europeans have de facto contributed to the construction of a hypothetical language which was never spoken by any existing community, as argued in the conclusions of Hanns Oertel's *Lectures on the Study of Language* (1901: 129) which discusses the views of Neogrammarians, Karl Brugmann, Hermann Osthoff and Paul Kretschmer. Speaking against any speculation beyond what facts strictly justified, later comparatists "drew attention away from the *Ursprache* as a supposed prehistoric reality to the data available in written records and in the spoken dialects of the present day" as stressed by Robins (1974 [1967]: 184). In consequence, the lack of any link between documented languages and their former bearers and the necessity of confirming a priori statements on the homogeneous nature of society has motivated researchers to investigate the speakers of living languages, which has resulted in the application of an ethnographical method.

This shift in attention to languages accessible to researchers with reference to the explanation of changes, contributed to the acceptance of a uniformitarian theory propagated in geology stating that geological processes operative in the remote past were no different from processes currently operative. Such an allusion to uniformitarianism, imported from natural sciences, led linguists, as one may read in Thomas Craig Christy's book *Uniformitarianism in Linguistics* (cited in Adamska-Sałaciak 1996: 16–17 and referred to in Collinge 1995a: 199 and 1995b: 205) to the belief that the evolution of languages proceeded in the same way both in the past and present times. In consequence, supporters of the belief of there being an uniformitarian evolution shifted the emphasis from the historical studies of documented dead languages to the recording and explanation of changes that resulted in the differentiation of living languages and their variability. Furthermore, the acceptance of the principles of uniformitarian evolution contributed to the epistemological shift from collectivism to individualism in the approach to the investigated object, that is, from the inquiry into the social existence form of language to the examination of the language of a communicating individual. The result of such an approach was that in fieldwork explorations, which started on the territory of Germany, and were then continued in various parts of Europe, the bearers of particular languages themselves had been selected as informants.

At the same time, the interest shift of researchers toward man became observable within the framework of naturalism as an investigative attitude. In terms of a biological concept of language as advocated by August Schleicher, man was viewed as a biological organism placed on the highest rung of the evolutionary ladder. Moreover, language was in Schleicher's view a physiological device for expressing notional contents and the relations between notions. On the whole, Schleicher considered language in itself also as a living organism brought up in a natural environment. Naturalist and materialist approaches to language have inspired, mainly in the late 1860s, subsequent explorations in the domain of the physiology of speech, the central nerve system and phonation organs as well as studies on the interdependencies between speech and brain, which have continued up until today.

Explorations in the domain of the geography of dialectal distributions of languages, initiated by Georg Wenker and Jules Gilliéron, have provided evidence that language changes are not explainable from the viewpoint of natural sciences exclusively through the principle of morphological analogy. Moreover, language changes are not regular, as Neogrammarians have claimed. Worthy of consideration also was the human factor with respect to territorial and historical conditionings. In particular, the marking off of the boundaries of particular linguistic phenomena by means of qualitative isoglosses on geographical maps have shown that sharp delimited dialectal zones do not exist. It appeared that every word had its own chain of speakers, and in fact quantitative isoglosses were important for showing the frequency of the occurrence of linguistic phenomena in the same territory. Confronting the methods of qualitative isoglosses with the methods of quantitative isoglosses as used in linguistic geography in order to register the boundaries between dialects and to check the instability of the output of their speakers under the influence of a standard variety was postulated by Witold Doroszewski in 1935.

Geographical methods of registering the instability of language norms which took place under the influence of a standard variety were not appreciated until after the 1960s with the development of urban sociolinguistics, especially by William Labov, as pointed out in *Trends in Linguistics* by Milka Ivić (1965 [1963]: §155). As we may learn from the discussion in Gerhard Helbig's monographic work on the history of modern linguistics (1982: 20ff), apart from traditional dialectology, geographism has given birth to cultural morphology, as developed by Rudolf Meringer and Theodor Frings in Germany, which focuses on the relationships of language to the artifacts of man with respect to cultural and historical-comparative aspect. Representatives of Neolinguistcs in Italy, or areal linguistics, *inter alia,* Matteo Giulio Bartoli and Giulio Bertoni (cited in Heinz 1978: 318–320), investigated also changes in the meaning of words or in the structure of language to establish the areas of language contacts, distinguishing the core and periphery with respect to the distribution of language phenomena. Geographical findings, discussed, *inter alia,* by N. E. Collinge (1995a: 200) and John J. Joseph, in an overview of the trends in 20th century linguistics (1995: 223), have revealed that, firstly, speaking collectives do not exist in their entirety; rather it is the case that what is available to empirical studies is only a communicating individual who appears to be an unstable speaker, and secondly, that the boundaries between territories inhabited by communicative communities (*Verkehrsgemeinschaften*) are fuzzy because particular words uniting separate chains of knowers do not overlap. The consequence of such studies was atomism of investigative objects.

One should recall here also the investigative perspectives of individualism, expressivism and mentalism which had characterized late 19th and early 20th century linguistic works. Their assumptions might best be summarized in three condensed statements coming from Andrzej Gawroński's treatise "La langue, sa nature et son origin" of 1927 (discussed in

Z. Wąsik's chapter on "The development of general linguistics in the history of the language sciences in Poland. Late 1860s – Late 1960s", 2001b: 29): Firstly, that language constitutes an articulated phonic form expressing the mental interior of man, which consists of intellectual elements, relating to the sense of words, their notional representations and emotional elements evoked by extra-linguistic factors as well as linguistic forms. Secondly, that language is subordinated to constant changes resultant from the changes in the mind of a particular human depending of how it expresses his needs, and that is always retarded as such to the development of human intelligence, because not all changes in the mind have to be reflected in language, when it functions as a relatively stable means of social communication. And finally, that language in the collective sense does not exist at all, since it is a kind of fiction, generalized on the basis of the social character of interpersonal communication, in which the habits and contents of communicating individuals underlie automatization. In contradiction, language as a personal property because of its changeability, underlies the will and the creativity of individuals.

2. The autonomy principle in the study of language as an abstract system of communicative means

When structuralism entered the scene of the prevailing perspectives, objectivism in the realm of investigative methods and factualism in the domain of investigated objects were postulated. The object of linguistic studies was understood in its initial phase as a system of static facts, which were realized in concrete speech (*parole*). The subject matter of the linguist's investigation was shaped by meaning-bearing elements and their relations to other elements within and outside of the system, which in turn was considered as an abstract social property (langue) (sociologism). Language in particular was considered not only as a system constituting a part of other systems of a higher range (systemism), but also as an autonomous system, which in turn could be divided into subsystems, viewed again as autonomous by researchers under a given gnoseological standpoint. The task of linguists was to identify and enumerate inventories of language entities, units and constructions. In opposition to the viewpoint of historical comparatists, some structuralists who adhered to the famous *dictum* stating that "language is a form and not a substance" of Ferdinand de Saussure from *Course in General Linguistics* (cf. 1959 [1916]: 122 or 1983 {1972 [1916]}: 120) postulated to investigate the systemic rules of language as a network of pure relational values.

Some others spoke in favor of searching for functions and the relational values of language facts, and not for the causes of their changes and evolution. In consequence, such an attitude meant a search for the functionally relevant properties of meaningful elements and their constituents and features. To state the matter more exactly, adherents of structuralist functionalism postulated a search for the invariants in the textual realization of the system of language in accordance with "the principle of abstractive relevance, as formulated by Karl Bühler, the Austrian psychologist and philosopher of language, for the theses of the Prague School phonology in his work *Sprachtheorie. Die Darstellungsfunktion der Sprache* (1965 [1934]: 25–28). Furthermore, it was referentialism in the structuralist view of language, which was derived from the thesis of the arbitrariness and conventionality of linguistic signs. The investigative object for structuralists constituted, on the one hand, the smallest element of distinctive function, namely phoneme, and the smallest element of semantic function, the morpheme, and, on the other hand, the basic entity of communicative function, namely the sentence. Dealing with the distribution and function of linguistic elements, linguists studied language in terms of synchrony. Even the diachronic studies conducted by Jerzy Kuryłowicz within the framework of structuralism were based on methods of the relative chronology of linguistic entities exposing productive and non-productive forms in the present state of a given language, as we may learn from the chapters of Z. Wąsik on "The development of general linguistics in the history of the language sciences in Poland" (2001b: 35) and Wojciech Smoczyński on "Jerzy Kuryłowicz as Indo-Europeanist and theorist of language" (2001: 266 mainly footnote 24).

The acceptance of the autonomy principle, derived from the opposition of autonomism to heteronomism, results in a depiction of language in which man is not seen as its master or creator, except with respect to the individual choice of a given style; he is rather treated as a language user. As such, a language is not a property of man. Rather it is the case that a communicating individual is said to acquire it through his own life. For language belongs to society as a whole and not to its particular members, and an individual, who speaks the language of his community, is usually corrected and experiences permanent social sanctions or normative pressures. As emphasized by William Downes in *Language and Society* (1998 [1984]: 33–35), when confronting the principle of autonomy and standardization with the heteronomous nature of dialects, it is prestigious to use standard, literary varieties, and dialects and vernaculars are generally stigmatized.

The idealization of norms within the framework of some sociological approaches to the existence form of language was accompanied by the psychological design of an ideal speaker-listener, who knows his language perfectly and is unaffected by errors as a member of a homogeneous community, having the same common experience and being unpolluted from any outside influences, as claimed by Noam Avram Chomsky in *Aspects of the Theory of Syntax* (1965: 3) and *Knowledge of Language* (1986: 17). As a result of the development of structuralist ideas, Saussureanism advocating a static view of language as a mental and social device was confronted with Chomskyanism which treated language as a dynamic property of an individual.

Against the background of European logocentrism and semiocentrism, language was identified with a set of facts and structures serving individuals as a tool (functionalism) for the purpose of interpersonal understanding. When it was said to manifest itself externally, it was considered, *inter alia,* by Michael Alexander Kirkwood Halliday in *Explorations in the Functions of Language* (1973: 105–108) as fulfilling threefold functions for the sake of man, textual function, i.e., communicative, interpersonal function – socializing, and ideational – cognitive. When it was eventually searched for in substantial realizations of human behavior, it was described in terms of stimulus and reaction (behaviorism). Being regarded as categories of abstract forms, which mediated between expression and content, it was studied independently of its realization in various substances (formalism).

However, adhering to the tradition of anthropocentrism in America, the followers of generativism claimed that individuals having at their disposal a finished number of constituents are able to understand and to create an unlimited number of new grammatically correct sentences never heard before. Here belongs also a relationist and mentalist view of language as a hierarchically layered code uniting the patterns of concatenated forms being organized at various levels of content and being related to its expression in linear sequences of sounds, which has been advocated within the framework of stratificationism by Sydney Macdonald Lamb in

Outline of Stratificational Grammar (1966) and popularized in David Lockwood's *An Introduction to Stratificational Linguistics* (1972: 144–49).

3. Homocentric and integrationist conceptions of the object of linguistic studies

A new path for a poststructuralist integrationism on a multidisciplinary basis, postulated, *inter alia,* by Roy Harris in his article devoted to "[t]he integrationist critique of orthodox linguistics" (1990: 74), has been prepared by functionalist contextualism and cultural anthropologism (Bronisław Malinowski, John Firth, and M. A. K. Halliday), as well as sociopragmatism (initiated by Ludwig Wittgenstein). A functional approach to utterances analyzed against the background of linguistic, social and cultural context has considered the fact that the structure of language is connected with the internal world model of a human individual. Apart from using structural methods derived from theoretical premises one observes in the investigative praxis a shift of emphasis from that of a generalized member of a society, treated as language bearer, to the individual user of a language constituting the property of a society. Moreover, researchers have recognized, as noticed in *General Linguistics* by Robins (1965 [1964]: 351), "that languages are not mere collections of labels or nomenclatures attached to preexisting bits and pieces of the human world, but that each speech community lives in a somewhat different world from that of others, and that these differences are both realized in parts of their cultures and revealed and maintained in parts of their languages".

What characterizes the poststructuralist linguistics is a homocentric conception of the object of linguistic studies, expressed through the perspectives of integrationism and communicativism, as discussed by Krzysztof Korżyk in a paper on language and grammar in the perspective of "communicativism" (1999) against the background of a "communicational grammar" and formulated programmatically, in an eclectic framework by K. Korżyk and Aleksy Awdiejew (2000) in their paper on "Communicative grammar: Toward a linguistic model of interpretative activity". Among the new domains of study, which do not necessarily exclude each other, one could name, for example, in the wider context: linguistic pragmatics, sociology of language, urban dialectology, sociolinguistics, the ethnography of speaking, text linguistics, conversational analysis, critical linguistics, embodied semantics, anthropological linguistics or linguistic anthropology, cognitive linguistics, and the like. Moreover, in a narrower context, the following disciplines should be placed: sociopragmatics, bilingualism and Creole studies, language planning, educational linguistics, etc. These are associated with philosophical, sociological and psychological frameworks of ethnomethodology, literary theory, postmodernism, hermeneutics, the ethnography of communication, biological and cultural cognitive sciences, discourse analysis, etc. Such language-, culture-, and organism-oriented studies concentrate essentially on the issues of multilingualism, multidiscoursivism and multiculturalism, as dealt with by the domains which concern themselves with the globalization of international communication.

The object of poststructuralist investigations have become texts embedded in social and cultural relations, discursive practices, speech acts, speech genres, as well as communicative events, dialogical utterances, conversations, belonging to the domain of the ethnography of communication, of which the basic terms, concepts and issues have been defined in *The Ethnography of Communication* by Muriel Saville-Troike (1982: 12–50). In the analysis of speech genres, researchers pay attention also to texts within texts, having the possibility of interpreting multi-layered utterances which depend on cultural, historical and personal experiences of communication participants.

Cognitive linguistics tends to present the individual's knowledge of the world in the form of mental reflections of concepts included in linguistic utterances, and consequently preserved in the social meanings of language entities, units and constructions. Enactionism, as opposed to mentalist constructivism, developed on the basis of the biological conceptions of Humberto R. Maturana, Francisco J. Varela, Evan Thomson, and Eleanor Rosch, *inter alia,* in *Anthropological Linguistics* by William A. Foley (1997: 8–11), postulates embodied semantics which claims that man as biological organism cognizes the reality through his senses. Hence, there is no objective meaning; it is embedded in the lived histories of human individuals. Such an understanding of meaning relegated to the subjective universe of an individual implies the necessity of negotiating the meanings of linguistic items by communication participants every time during the process of conversation. The linguistic communication of man is studied also along with axiological issues, ideology, values, ethics, and subjective needs realized through verbal understanding in spoken and written communication.

Taking into account the various diverse poststructuralist conceptions of language, both in the domain of its ontological status and the methods of its investigation, one can reconstruct a picture of man who speaks various languages, and who utilizes various registers, dialects and vernaculars of the same language. Man as a representative of a hominid species is characterized by biological and cultural diversity, and members of ethnic and national communities are also diversified and stratified socially as members of discursive communities of topic-, domain- and/or profession-related character, etc. Man, as a speaker, is not considered ideal, because he makes observable errors, as revealed in discussions and criticism of the idea of homogeneous and heterogeneous, monolingual and multilingual linguistic communities by Alessandro Duranti in *Linguistic Anthropology* (2000 [1997]: 72–76]) and Victor H. Yngve in *From Grammar to Science* (1996: 69). Such a speaker is subject to social pressures referring to correct language usage, and so on. Accordingly, language belongs to the shared means of interpersonal communication in which individual participants realize their tasks, intentions, or plans, either directly or indirectly, through their verbal and nonverbal behavior in speech acts and communicative events.

Within a small-group community, interpersonal communication is not only interactional but also transactional by nature as pointed out, *inter alia,* by Joseph A. DeVito in *The Interpersonal Communication Book* (1976: 101–127), discussing the principles of transactional analysis popularized by Eric Berne and Thomas Harris. Interactional relations occur in communicative events when there is a reciprocal action, and a mutual exchange of messages between sources (authors, senders) and destinations (addressees, receivers), that is, every participant of interpersonal communication functions in exchangeable roles, sending or receiving messages. The notion of transaction presupposes that people are not always the same in diverse social relations and communicative situations. They alter and adapt to each other under the influence of social, cultural, physical and personal conditionings of communication. One can therefore describe a communicating individual as a social actor playing certain roles, striving to achieve a certain intended goal as a member of institutional communities organized around a network of collective requirements, ideology and morality standards, in accordance with the investigative tasks of cultural anthropology, as summarized by Allessandro Duranti in *Linguistic Anthropology* (2000 [1997]: 3). Through words, he can change the reality which surrounds him; he may either loose or gain the understanding of his interlocutors. Furthermore, he can anticipate or evoke certain presumed behaviors. The identification of language-use with the communicative acts of speech leads in turn to reflections over, e.g., the ways of how to realize certain ideas, satisfy certain values, and how to fulfill the various needs of interlocutors.

To sum up, we recall the statement of the sociolinguist William Downes who states that "Linguists investigating actual speech are interested ultimately in individual speakers. But they are studying them also, inevitably, as members of groups" (1998 [1984]: 105). There are however representatives of the sciences of language, who name speakers/listeners or communicating individuals, and, in general, people, in the first place as the real object of linguistic studies. Among them are Franciszek Krzysztof Grucza and Victor Huse Yngve.

Franciszek Grucza has developed a human-centered approach to the object of linguistic studies with special reference to glottodidactics starting from the late 1970s (cf. Grucza 1979, 1983a, 1983b, and 1994). His ideas became widely known through his book devoted to the issues of metalinguistics – linguistics, its subject matter, and applied linguistics (1983a). For Grucza (1983a: 284–297) language is inaccessible to immediate observation and only the people who speak it and their utterances exist as the real objects of linguistic study. From this perspective, language constitutes a mental property of people in such a way that linguistic rules and the abilities of how to use them are internalized in the consciousness of speaking and hearing individuals. In Grucza's view (1983a: 418), language speakers/hearers, while making use of that ability, develop their skills in communicating with each other by means of utterances that are the material bearers of semantic functions. Grucza's (1983a: 359–389) understanding of the object of linguistic studies in terms of the functional properties of human mind was postulated, in consequence, for defining the tasks of glottodidactics (1983a: 440). Applied linguists were advised to study people with respect to their linguistic and communicative aptitudes, in order to find answers to the question regarding what should be done so that students learn foreign languages with fewer errors, thereby developing their verbal skills efficiently.

Victor H. Yngve has formulated a slightly similar conception concerning the tenet that only the accessible objects of linguistic studies are people. The foundations of this conception may be dated to 1970s (1969 1975abc, 1986, 1996), nevertheless representative for its recent applications is the book *From Grammar to Science* (Yngve 1996: chapters 7 and 9 for background, 10 and 14 for the basic theory, and chapters 15 to 18 for illustrative purposes). In a concise form, it is available though his programmatic articles "Issues in hard-science linguistics" (Yngve 2004a) "An introduction to hard-science linguistics" (Yngve 2004b) and "The conduct of hard-science research" (Yngve 2004c).

However, for Yngve people are relevant with respect to the linguistic properties of individuals and groups, which enable them to communicate with other individuals. Moreover, sound waves, and other kinds of energy or props that unite individuals into temporary and long-lasting linkages are said to belong to the world of truly physical objects bearing no information or meaning at all. As one may conclude from Yngve's human-centered view of verbal and nonverbal communication, the objectives of hard-science linguistics are at least twofold. Firstly, scientists are expected to find answers to the questions of what the communicational tasks of people are and how they are realized (as it has been revealed in E. Wąsik's 2004 article "Describing Frisian communities in terms of human linguistics"). Furthermore, they have to find out what kind of dynamic relations constitute the linguistic structure of individuals and people with reference to the properties that determine their social roles as participants of communication realized in verbal and nonverbal behavior.

From the perspective of the broadly understood non-linguistic sciences of language, linguistics proper and human communication, the framework of human-centered linguistics, elaborated as hard-science linguistics by Yngve, which used to be called also "human linguistics", could be discussed along with the recently established disciplines of communicology and social sciences against the background of the semiotic-grammatical tradition of ancient linguistics (cf. Yngve and Wąsik 2004a, "The riches of the new world" and 2004b "Coping with cultural differences"). Traditional linguistics, reconstituted in terms of hard sciences and human linguistics, could deal with the explorations of the changeable nature of linguistic structures (hitherto included into the realm of linguistic semiotics) originating in dynamic links that unite people though verbal discourses and speech events.

Chapter Two

An inquiry into the functional view of language as a tool of communication or a property of task-realizing communities

Being concerned with functionalism as applied to the sciences of language, we have to clarify, firstly, its place among other investigative perspectives chosen by the scientist in order to highlight or to view the aspects of their objects of study found in their investigative domain. Secondly, we have to give attention to its constitutive category selected for delimiting the boundary of the investigative domain, namely, the notion of function, which should determine one of the properties of the investigated object. To do so, we will finally have to check the uses of the term *function* in functional statements.

1. The interdisciplinary background of functionalism as an investigative perspective

Depending on the nature of functional objects and the respective understandings of function derived from different frames of reference, there are multifarious faces of functionalism as cognitive attitudes or cognitive standpoints of scientists. Generally speaking, there are two sources of linguistic functionalism: the first one is connected with the search for functional invariance in language, and the second with the search for various uses of language. To inquire about their provenance we have to make a survey of functionalist perspectives which were widespread at the end of 19th and the second quarter of 20th centuries, on the one hand, in such disciplines as biology, psychology, cultural anthropology, and, on the other, in architecture and art. As a result of interference between the methodologies of applied arts and architecture and the behavioral sciences, biology and psychology, it is possible to distinguish in the domain studied by social sciences, cultural anthropology and sociology two ways of functional thinking, and in effect two kinds of functionalist perspectives.

The first kind of functional approach to language arose in the "climate" of an essential utilitarian instrumentalism which dominated in philosophy, literary studies, architecture and applied arts in the 1920s and 1930s. But it borrowed some categories and notional distinctions partly from the second kind of functionalism, namely the benefactive-contributory organicism prevailing in biology, psychology, cultural anthropology and sociology. Organicists strove to exhibit the dispositional properties of an investigated object satisfying the requirements of a given environment or contributing to the proper activity of a given living system. The followers of instrumentalist views sought to reveal such properties on the principle of 'abstractive' relevance, thereby aiming at the detection of functional invariants within the concrete substance of an investigated object playing a serviceable role for its user.

One speaks about an instrumentalist functionalism (or a functionalist instrumentalism) when the fulfillment of a function by a given element depends upon the purposes which are externally determined, that is, when the element, regarded as functional, is used as a means for accomplishing a certain task or purpose of an acting subject. However, when the homeostasis, i.e., the tendency to maintain internal stability of a living system, as a self-sustaining being, is exposed due to the normal functioning of its parts, then the perspective in question is called an organicist functionalism (or functionalist organicism).

It should be noted that instrumentalist functionalism, had developed in applied arts, and urban architecture in the 20s and 30s of the 20th century, where the attention was paid to the abstractively relevant features inherent in the structure of elements playing a serviceable role with regard to human needs and social requirements. Through the mediation of urbanist constructivism this movement turned into a kind of functionalist structuralism, which, having accepted the principle of abstractive relevance, aimed at separating in the elements and/or structures, things or events, what is typical and general from what is accidental and individual. In other words, an attemptt to determine what is relevant and what is irrelevant, with regard to their primary functions.

1.1. Functionalism searching for invariant features based on the principle of abstractive relevance

An instrumentalist functionalism exposing the features of elements that are abstractively relevant for playing a serviceable role in relation to their users or makers was popularized in the program of the Prague Linguistic School, particularly in the context of the distinction between phonetics and phonology. It arose undoubtedly in the "climate of opinion", prevailing at that time in Austria, Switzerland and then in Germany after the World War I, which promoted the perspectives of purpose-and-need-oriented rationalism in architecture, utility products, and environmental urbanist constructivism. As a mode of thinking and a tendency among intellectuals and architects, functionalism was considered as being opposed to expressionism. Its propagators postulated to consider the needs of average people while producing the utility goods which are rationally conformed with/to the purposes of everyday life, and which are not falsified by abundant ornament and excessive useless form, as exposed by Adolf Loos, an Austrian architect, in his manifesto "Ornament und Verbrechen" (1981 /1964/ {1962 [1908]}; for details, references and quotations see in Horn 1968: 110–11, and 143). This rationalist principle demanded that architecture and applied arts should reflect the pure relationships between man and his environment while taking account of his biological, social and culture-creative nature.

As Walter Gropius, the founder of the Bauhaus School in Weimar and then in Dessau stated – the concept of rationality is based on the function of what is adequate in relation to purpose, and the beauty is the mirror of what grows as a result of appropriate use for specified purposes.

The methodology of purpose-need-oriented rationalism appealed to the following principles: (1) the Ockham-razor-principle, implying the fight with Kitsch and ornament, imitation and stylization, (2) the ergonomics principle, stating that design of devices, systems, and physical working conditions should be coordinated with the capacities and requirements of the worker, (3) the economy principle, involving the requirement analysis, and (4) the testing of prototypes (cf. Berndt, Lorenzer and Horn 1968, separate articles; Moles 1971: chapter IX, for detailed information and references).

1.2. Functionalism exposing teleological or etiological explanations of change and adaptive processes

The notion of function in biology was connected with functional statements or functional explanations, ascribing directly or descriptively the activity or role of certain organs for the attainment of final goals of an organism as a whole. The goals of living systems were seen in (1) striving towards the preservation of species (reproduction, procreation), (2) the maintenance of constant forms of existence and survival (homeostasis), and (3) the steady development towards the natural end by providing the energy in the physical and chemical processes of consumption, utilization and, finally, excrementation, i.e., waste matter discharged from the body (metabolism).

Followers of naturalist evolutionism, based on Charles Robert Darwin's view of the development of living species, stressed the role of 'natural selection' in the 'adaptation' of individual organisms to conditions encountered in their ecology. Trying to detect the character of natural changes, they asked for functional explanations as to why certain organs disappeared or were modified in form by nature, being functionally exploited or not. Independently of philosophical debates on the notion of function and purposeful activities, – as, for example, in the case of a spider who produces a net to catch a fly, or a fisherman who produces a net to catch fruits of the see – the anthropomorphic approach to the teleological explanation in biology followed the principle that, all organisms constitute goal-directed autonomous systems that possess their own subjective universe of needs and their own functional circles. Such an approach led, in consequence, to the statements that some requirements have to be fulfilled in order to sustain their reproductive, homeostatic, and metabolistic potentials of organisms.

Functionalism in psychology appeared towards the end of the 19th century as a set of opinions opposed to structuralism. Psychologist functionalism stressed the role of the mind as a biological tool that is utilized in the adaptation of human individuals to varying requirements of the environment, i.e., the role of thinking in the formation and optimization of the living conditions. Functionalism developed in the investigative domain of psychology paid attention rather to mental activities than to mental contents, aiming at an explanation as to what way judgments, perceptions, feelings, desires and other dynamic acts of thinking are helpful to man to resolve his problems which he encountered in everyday life.

Although psychologist functionalism was not uniform as a trend represented, *inter alia,* by John Dewey's (1896) "The reflex arc concept in psychology"; James Rowland Angell's (1904) *An introductory study of the structures and functions of human consciousness,* and (1907) *The province of functional psychology,* as well as Franz Brentano's (1874) *Psychologie vom Empririschen Standpunkt,* recapitulated in later years by Hervey Carr's (1925) *Psychology, a study of mental activity,* some of its principles had been summarized in the following statements: (1) psychology is to be concerned rather with mental functions than with contents, (2) psychological functions are interpreted as human adaptations to the environment, (3) psychology should be utilitarian, i.e., allowing practical applications, (4) mental functions constitute a part of activity of the whole world, which embraces both mental and physical activity, (5) psychology is closely related to biology, so that the understanding of anatomy and physiology is helpful in the understanding of mental activity.

1.3. Functionalism referring to the dispositional properties of objects satisfying subjective needs of organisms and the systemic requirements of culture and society

A functional model of conducting studies in the domain of cultural anthropology appeared in the year 1922. It was marked by the publication of two works, *Argonauts of the Western Pacific* by Bronisław Malinowski and *The Andaman Islanders* by Alfred Reginald Radcliffe-Brown. Malinowski stressed the study of culture as a teleological system constituting an instrumental apparatus which enables the man to adapt to his natural and social environment. In Malinowski's view, every part of the culture was to be considered as a means to an end. The functionalism of Radcliffe-Brown, in turn, was based on the notion of social function analogical to the organicist notion developed in biological sciences. For Radcliffe-Brown (cf. as summarized in Merton (1957 [1949]): 22): the function of a particular usage, as a socially standardized mode of activity, or mode of thought, was its contribution it makes to the existence and continuity of the total social system.

Close to the views of Malinowski, one should also discuss Talcott Parsons (1949, and others 1967 [1951]), the sociologist who combined the achievements of the psychology of personality with the problems of human needs within the framework of functionalism. In Parsons's view, social systems constituted roles of particular personalities determined by cultural patterns, as systems of norms, standards, rules, values, ideas, beliefs, etc. These roles are internalized by individual persons and institutionalized by social groups. The functionalistic conception of sociology according to Parsons stresses the motivational character of the action of individual members of societies oriented toward the achievements of particular goals and organized around certain roles within a particular social system determined by individual and social needs, labeled as functional requirements or imperatives, such as: (1) adaptation, (2) attainment of a goal, (3) integration, (4) sustainment of a cultural pattern adjusted to individual and collective needs.

Another interpretation of functionalism was developed by Robert King Merton (1957 [1949]) who stressed the importance of functional analysis in the study of society as a system of particular elements. Merton postulated to investigate functional consequences in the interpretation of cultural products and behavior which contribute to the survival, maintenance, and compactness of the whole society. From the viewpoint of consequences Merton distinguished functional alternatives, functional equivalents, and functional substitutes, being aware that some cultural forms and institutional activities, some patterns or practices, can be functional for one social group and dysfunctional for the other. Functions and dysfunctions as consequences for the society depend, in Merton's view, on the one hand, upon subjective dispositions (needs, interests, motives and purposes) of speaking individuals and communities, and objective functional consequences not limited to conscious and explicit purposes, on the other. Guided by the premise that all cultural artifacts fulfill social functions, Merton arrived at the conclusion that they can be detected as either manifest or latent functions. Assuming that there are not only consciously motivated consequences which are intended and expected by social subjects (having a subjective aim-in-view) but also unintended and unrecognized consequences as caused by incidental circumstances, Merton equated the function in the teleological sense with purposes and with causes and effects in the etiological sense.

The debates over the essence of a functional approach to society and its culture within the framework of biological organicism were summarized by philosophers of science, such as Ernst Nagel and Carl Georg

Hempel. Nagel (1967 ([1956]) proposed "A formalization of functionalism" in replacing functional statements by unified formulae using synonymous descriptive terms, as, e.g., "The function of X is Y" by the statement: "X is a necessary condition for Y", and Hempel ([1959] 1965) confronted functional studies in biology and psychology with functionalism as an investigative perspective in sociology on the basis of differences which ought to be observed between functional statements and teleological statements. In Hempel's view the idea of universal functionalism seems to be untenable as long as scientists are interested in searching for self-regulatory aspects of social and other systems while comparing them to the biological organisms that strive for homeostatic and regenerative or reproductive processes.

In turn, the reflections of these philosophical debates on the nature of functionalism in biology and sociology had found its best expression in the philosophical works of Larry Wright (1976) and Andrew Woodfield (1976) devoted to the distinction between etiology and teleology. It is worh comparing the descriptive statements used by the latter to avoid the use of the term *function* while replacing it with the paraphrase "in order to" (Woodfield 1976: 19, 32, interpretation is mine: EW):

(1) The fisherman makes the net in order to catch fish (i.e., food) – teleology.
(2) The spider weaves the web in order to catch flies (i.e., food) – etiology.
(3) The man ran in order to catch the train – teleology.
(4) The cat opened the door in order to get the cream – etiology.
(3) The thermostat turned the heater off in order to stop the water going above a certain temperature – etiology.
(4) Witchcraft persecutions occur among the Navaho in order to lower intra-group hostility – teleology.
(5) The heart beats in order to circulate the blood – etiology.
(6) Knives have blades in order that they may cut – etiology.

Following the reasoning of philosophers it has been observed that the understandings of function, for example, in biology and socio-cultural anthropology are not the same when it is ascribed to natural body parts or processes fulfilling the goals of an organism or cultural artifacts as purposeful products satisfying the needs and requirements of its producers. The distinction between etiology, aiming at the explanation of causes, causation or causality, and teleology, searching for the evidences of design or purpose in nature and culture, appears to be of crucial importance. In biology, the term *function* is used in both the behavioral sense, that is, in terms of a goal-directed act, and in the vitalist sense, that is, in relation to the role of an organ or organismal processes within an organism as a whole. Behaviorists claim that human psychology can be accurately examined through the analysis of objectively observable and quantifiable behavioral events as they occur in the world of animals; but the same cannot be said about the subjective mental states. The followers of vitalism, in turn, attribute the viability of a living organism to a vital principle distinct from the physical and chemical processes of life. Adapted from biology to the field of socio-cultural anthropology, the functional way of reasoning, as a cognitive attitude, is based on the justification or explanation of the *raison d'être* of a given element with regard to its function within a given society or culture treated as organism, or on the search for functions, which a given element fulfils with regard to its source or destination, with regard to an acting subject or with regard to a frame of reference, in which it is situated.

2. Towards a paraphrasing explanation of the term *function*

Asking for the real definition[1] of the term *function* it is not sufficient to find *its genus proximum* in synonyms of action or activity proper to a person, thing, or institution; the purpose for which something is designed or exists, or the synonyms of purpose, role, activity, operation, job, business, task, duty; capacity, power, faculty; province, sphere, office, niche, place, scope, range, field; concern, objective, *raison d'être*, and verbal usage, meaning to perform a specified action or activity; work; operate – as specified in dictionaries. It is also necessary to find contextual *differentia specifica* exposing additional nominal or verbal equivalents not known in the stylistics of everyday speech, as, for example, utility, status, means and end, result, outcome, achievement, exponent, value, good, meaning, sense, effect, operation, disposition, etc. depending on the meaning of the context in which the proposition is placed: substituting as such the synonyms of purpose, role, activity, operation, job, business,

---

[1] In search of the usage of the term *function* out of our concern should be a non-functionalistic notion connected with a ceremonious public or social gathering or occasion, being synonymous to fete, gala, affair, festivity, party, entertainment, reception, soiree; dinner party, feast, banquet; ceremony, occasion.

task, duty; capacity, power, faculty; province, sphere, office, niche, place, scope, range, field; concern, objective, *raison d'être*, and verbal usage, meaning to perform a specified action or activity, to work or to operate.

Trying to figure out the notion of function from functional statements, one can consider, with reference to philosophical debates,[2] such formalized paraphrases in which X is an object functioning, having, or performing a function F and S is a subject, a frame of reference, or a system in which or for the sake of which the function is being performed: In some understandings of function it is not necessary to speak about the subject for the sake of whom an object functions. Some others stand very close to needs / requirements and some to ends / results which are not intended by subjects. Still others mention the role of an object X for the objects X of other kinds or also the X's functional consequences for certain frames of reference. Accordingly, one may distinguish, at least, the following meaning types of function:

(1) Semantic significance: X has a meaning for S by performing F; The sense of X is to perform F for S, in S.
(2) Instrumental teleology: X is a tool (a means) for performing F; X performing F is a means to achieve S's goal, or a purpose intended by S.
(3) Consequential etiology: X performing F causes another X. The performance of F by X is a cause of another X for S, in S.
(4) Exponential indexicality: X is an exponent of F in S; The F of X is to indicate its properties for S.
(5) Constitutive componentiality: X performs F as a constituent of S.
(6) Utilitarian potentiality: X is useful for S by performing F.
(7) Performative assistance: The F of X is to serve S; X acts as a servant by doing F; X is in the service of S (or) X works for S.
(8) Operational pragmaticity: The performance of F by X is an action on behalf of S; X acts in favor of S by performing F.
(9) Axiological assessment: The F of X has a value for S.
(10) Resultant finality: The F of X is to attain an end, a result.

---

2   Summarized in a paper under the title "Detecting the meaning of function from the paraphrases of functional statements" delivered by Zdzisław Wąsik at the 40th Annual Meeting of the Societas Linguistica Europaea (SLE 2007): *Functionalism in Linguistics*, 29 August–1 September 2007. University of Joensuu, Finland.

(11) Dispositional efficiency: X has a capacity for producing a desired result or effect by performing F.
(12) Attributable obligatoriness: S or X by doing F performs the duties in S or for S.
(13) Benefactive contributiveness: The F of X is to contribute to the good of S; X is good for S by performing F; X endorses the well-being of S by doing F.
(14) Gratificational adequacy: The F of X is to gratify (to satisfy, to please, to answer) S's requirements, needs, wants, expectations, appetites, desires, inclinations, or feelings, etc. By performing F, X is sufficient or adequate as for the needs or purposes of S.
(15) Ecological adaptiveness: X performs F in a certain place or for the sake of a certain environment.

With reference to semantic inquiries into the understandings of the term *function*, one can state that the differences in a functional description of language characterizing various schools of linguistic thought, have been expressed either in purpose-oriented (teleological) or cause-and-effect-oriented (etiological) terms. They have resulted from the oscillation between the understandings of function: (1) as an exponent of a certain role or (2) a relation between certain elements that play a serviceable role and of function which is synonymous to (3) purposes or (4) the ends of certain activities, intended or not intended. Thus, functionalism as an investigative perspective in linguistics appears to refer, in general, to such notional and methodological frameworks: (i) which take a functional view on the nature of language, (ii) which attach primary importance to functional relations at different levels in the organization of language; (iii) which mean the practical method applicable to the analysis of diverse aspects of language and language use, or (iv) which specify "functional realms" being encoded in human language.

3. A tool-oriented view of language functions realized by verbal means as one of the constituents of a communication scheme

Whilst generalizing the various uses of the terms *function, functional explanation* and *functional relevance* in linguistics, one may notice that linguists have either applied the principles of functionalism in relation to the whole system of verbal means of communication or to its particular elements. Some share the belief that there is only one function of language, which can be derived from the assumption about the serviceable role of verbal means, and some others accept as true that one can enumerate a number of principal functions of language, bearing in mind that the products or acts of speaking and understanding activities can realize various purposes of language speakers or users. Not to be omitted are those researchers who place among language functions also observable consequences, as the results of the use of language. Thus, on the opposite poles are, firstly, the practitioners of linguistic sciences, who are interested in linguistically and communicatively relevant features of verbal means, and, secondly, the theoreticians who pay attention to the serviceable role of language in the fulfillment of tasks (aims or purposes) of communicating people by the utilization of verbal means. In summary, it is impossible to speak about the functions of language in general. The researcher has to choose such an understanding of language, which exposes its aspects and constituents, its manifestation forms and the modes of its existence.

3.1. Alan Gardiner's definition of function as a performance ability of language products to act in subservience to certain aims in the view of language users

The term *function* as a synonym of performance, and especially the result of certain activity appeared in the works of Alan Henderson Gardiner, a British representative of structural linguistics. Gardiner noticed that the term *function* is used to denote a certain capacity to act or the possession of an entitlement to perform something, i.e., a certain duty in subservience to the aim or use of an acting agent. Cf. the following quotation:

> Etymologically, 'function' is only a rather grand synonym of 'performance', but it is often used in a peculiar way to designate the capacity in which something acts in subservience to a certain aim. Thus a nail driven into the wall can function as, or have the function of, a peg to hang one's hat on. Two conditions govern this use of the word: firstly, that some particular type of service should be named to indicate the capacity in which the functioner acts; and secondly, that the aim or purpose subserved should be that of a human employer. These notions reappear in the linguistic use of the term where it has reference to the results achieved in the course

of a particular act of speech. In such an act the speaker's aim is to draw attention to something, and the words are, as it were, his functionaries whose office it is to present the thing-meant as possessing some particular formal character. However, grammarians seldom avail themselves of the term 'function' in reference to the general duties of a word in the fulfillment of its inherent 'form' or capacity. The term comes into play only when details of the work done have to be described, or when the word is found doing work which is not, properly considered, its own." Gardiner (1932: 141–142).

3.2. Karl Bühler's representational function and the principle of abstractive relevance in the instrumental model of language

Karl Bühler cited Gardiner in his work *Theory of language: The representational function of language* ([1934. *Sprachtheorie. Die Darstellungsfunktion der Sprache*]) referring to the concept of language as a tool (*Organon*) derived from *Kratylos* (Κρατύλος), the dialogical work of Plato (427–347 B. C.). Without any doubt, Bühler's idea of detaching functional from non-functional properties of this tool came from the idea of a purpose-oriented rationality (*Zweckrationalität*) prevailing in the Austrian art and architecture at his times.

The contribution of Bühler to the achievements of the Prague school of linguistics is known due to his communication scheme called *Organon-Modell der Sprache*. However, there are different interpretations of his view of language functions, which must be cleared up:

Firstly, in his Organon-Modell, Bühler distinguished four concrete constituents: sender (*Sender*), sign (*Zeichen*), receiver (*Empfänger*), and extra-semiotic reality, meant as things and states of affair (*Gegenstände und Sachverhalte*). Among them the sign, was considered as the main constituent standing in relation to the other three.

Secondly, the communicative function of language (*Mitteilungsfunktion der Sprache*) was realized, according to Bühler (1965: 28), through a threefold performance of interconnected functions of the linguistic sign: in relation to a sender, it is a symptom (*Symptom*) having an expressive function, in relation to a receiver, it is an appeal signal (*Appell*) being able to perform an impressive function, and, in relation to things and states of affair, it is a symbol (*Symbol*) being endowed with the ability to perform a representational function. Only the representational function, called later the semantic function, was treated by Bühler as principal for the function of language, namely, *die Darstellungsfunktion der Sprache*.

And finally, Bühler's aim was not to launch the idea of language functions but to expose the fact that some features of verbal means are functionally relevant and some are functionally irrelevant. These features are to be separated from each other through the so-called principle of 'abstracting' relevance (*das Prinzip der abstraktiven Relevanz*). In consequence, Bühler distinguished also redundant features of the sign among functionally relevant features, which are omitted in the process of faster speaking, but may be supplemented through apperception (*apperzeptive Ergänzung*), i.e., through the memorized traces of experience.

Bühler's deliberations on the functions of the linguistic sign have inspired a wide range of scholars in Europe. Among the main representatives in Poland to be mentioned are Tadeusz Milewski (1973 [1965] cf. also 2004 [1965]), and Leon Zawadowski (1966, 1975). Under the influence of Bühler's theoretical proposals Milewski (1965: 49–50; 2004: 37–38, cf. mainly Milewski 1965: 49; 2004: 37) distinguished in the same manner the three basic functions of human speech. Along with the semantic function, based on the reference of speech products to extralinguistic reality, he distinguished two other functions resultant of speaking and hearing activities, which, in fact, are non-semantic in reality: an expressive function, when the speaker is characterized by the features of his voice, and an impressive function, when the features of the speaker's voice evoke the feelings or reactions of a receiver. The expressive and impressive functions cannot be interpreted on the same level of goal-oriented phenomena. The first one, namely, the expressive function, is non-teleological, i.e., etiological, as far as the voice characteristics are not pre-planned intentionally by a speaker, or do not depend upon his will, and the second, the impressive, is teleological while being dependent on the purposeful activity of the speaker with the aim in view of making an impact upon the emotional state and behavior of the receiver.

The Prague School legacy to distinguish the functional from non-functional features of speech products, called texts as the substantial realization of language having an abstract and social character, constituted the core of the epistemological reasoning in Leon Zawadowski's linguistic theory of language (1966: 130–160). Having accepted the definition of language as a set of types of text elements which serve its users to speak about extratextual reality, he assumed that the functions of human speech should be searched for in the domain of texts realized on the basis of a unilateral implication as semantic and categorial correspondances between verbal means and their referents. Moreover, Zawadowski postulated to regard the representational function as the principal semantic function of language with respect to its individual user. Another type of semantic function is, for him, the communicative function, carried out in the speech event when both interacting individuals, the speaker and the hearer, know what the text represents in accordance with mutually accepted conventions. Following the principle of abstracting relevance Zawadowski was of the opinion that the semantic function of texts is played by its minimal set of differential features. Above and beyond, the common mass of text features constituting the domain of social norm and the maximal set of features characterizing individual communication participants have been regarded as engaged in the fulfillment of non-semantic, expressive and impressive functions.

3.3. A critical review of Roman Jakobson's eclectic distinction of six functions of language as parallel to six communication constituents

What Roman Jakobson did in "Linguistics and poetics" (1960) was, at the first sight, that he added only two additional constituents to Bühler's fourfold communication scheme. Considering the six constituents altogether, such as: sender, receiver, context, message, contact and code, one may interpret Jakobson's way of reasoning as follows: A sender transmits a message to a receiver, as far as this message refers to a certain context. Communicational requirements demand that the speaker and receiver have a common channel, which enables them to maintain physical contact, and a common code, which allows them to understand each other.

With reference to the six constituents of communication, Jakobson distinguished six parallel functions of language, namely *emotive, conative, referential, phatic, poetic*, and *metalingual*. Conforming with a critical assessment of Jakobson's functions made by Antoni Furdal (1990 [1977]: 50ff), one has to recall that, according to Bühler, it is the message, which stands, as one of the four constituents, in a functional relationship to the remaining three others. For Jakobson, however, each of the functions is conditioned by each of the constituents of communication; the emotive function is oriented towards the sender, conative – towards the receiver, and referential – towards the context. Besides, the metalingual function is resultant from its orientation towards the code, and the so-called phatic function – towards the contact, and finally the poetic function – towards the message itself.

In reality, the first three functions distinguished by Jakobson, emotive, conative and referential are only superficially similar to Bühler's expressive, impressive and semantic-representational functions. Bühler treated them altogether as a threefold function performed by one and the same sign. In Bühler's *Organon-Modell*, the "symptom-indicating", i.e., expressive function was understood as a non-intentional expression of the sender's features as perceived by the receiver; whereas Jakobon's emotive function was rendered as *nota bene* an intended expression of feelings and reactions. The second type of function, namely, the "appeal-signaling" in Bühler's and the "conative" in Jakobson descriptions, is usually rendered as the impressive function. However, the term *conative*, (derived from the Latin *conatio* 'an effort' = *cona (ri)* 'to try'; cf. also English *conation, conational, conative*), additionally denotes those aspects of mental life, which have to do with purposive behavior, including desiring, resolving, and striving. As far as Bühler's symbolic-representational and Jakobson's referential functions are concerned, they are not identical. The symbolic function pertains to observational meaning-bearers; in turn, the referential function, entails also, apart from the observational semantic functions, some other concluded cognitive functions.

The three additional language functions, distinguished by Jakobson, as metalingual, phatic and poetic, are also questionable. Both the metalingual function coming into existence when the language constitutes the extra-semiotic reality of communication and the phatic function connected with the initiation and fostering of conversational contact are to be regarded as instances of semantic-representational functions. In the case of the so-called metalingual function of speech products, one should rather speak about the meta-designational nature of speaking where the speakers form meta-signs as meta-statements about the statements of a given language.[3] As to the phatic function, it should be added that Jakobson's idea

---

3   As pointed out by Zdzisław Wąsik (2003: 60), the term *metadesignation* has a long tradition, which goes back to antiquity where it is related to the idea of *metalanguage*. Christoph Hubig (in Sebeok 1986, 529–31, entry: *meta-*) and Jacek Juliusz Jadacki (in Sebeok 1986, 445; entry: *metalanguage*) provide evidence that it was Stanisław Leśniewski, a Polish logician and philosopher, who introduced for the first time to modern logic the distinction between "language" and "metalanguage".

of *phatic-ity* borrowed from Bronisław Malinowski (1923) does not conform with the intention of its author. Malinowski has coined the term *phatic communion*, in order to describe such a type of speaking in which intimate bonds of unity are formed solely through the exchange of words, that is, by speaking only, the purpose of which is, in the first instance, to establish and to sustain the existence of social linkages among participants of communication. Jakobson (1960), however, has used Malinowski's term to describe the situation, in which the channel between interlocutors is open. From that time on, the adjective *phatic* (derivable from the Gk *phat (ós)* spoken, capable of being spoken, as a verbal adjective of *phánai* to speak), is usually perceived as denoting a kind of speech used to express or create an atmosphere of shared feelings, goodwill, or sociability rather than to impart information.

One should, therefore, agree with Furdal's statement that, in dependence from potential interpretation of messages, the number of subsequent language functions could be set up *ad infinitum*, depending upon features that evoke attention, reactions or feelings of the receivers. Therefore, it is better to speak about the poetic function of language in a special artistic usage of verbal means. In Jakobson's understanding such a function is fulfilled, when the message finds itself in the center of attention for the sake of or with regard to itself.

To understand the idea of poetic function, it might be important to explain it against the background of the esthetic function of language introduced by Jan Mukařovský (1970 [1936]). The esthetic function, explainable in terms of the so-called *autotelic function*, is connected with social norms and values, which are realized and sensed individually.

By autotelic function Mukařovský meant that estheticity or estheticism is to be understood as a purpose for and in itself. Even so, the empirically oriented practitioners of language sciences may firstly ask, to start with, who the subject is who determines the existence of such an esthetic or poetic function. Secondly, they may ponder whether the poetic function is individually or collectively estimated. Thirdly, the question arises who is entitled or enabled to assess what esthetic is, the receiver

---

The term metalangue was put forward (in oral presentations of 1931 and thereafter widely accessible publications of 1935 and then 1956) by Alfred Tarski with reference to the area of the theory of truth. A separate study to "The problem of metalanguage in linguistic historiography" was discussed by Ernst Friderik Konrad Koerner. For further information and detailed references see Koerner 1993.

or the creator; and finally, it must be also considered whether the art has any subject at all.

4. A survey of functions attributed to language as a socio-psychological property of communicating individuals

4.1. Socio-pragmatic and cognitive perspectives on the intersubjective functions of language in M. A. K. Halliday's and Robin P. Fawcett's depictions

A linguist who has confined himself to the understanding of function connected with services of a certain object fulfilling a certain aim or purpose in-view of a certain actor is not willing to equate Michael Alexander Kirkwood Halliday's functions of language (1972 [1970]: 143, discussed also in Kress (ed.) 1976: Introduction xix.), as *ideational, interpersonal,* and *textual,* with Karl Bühler's functions of the linguistic sign, such as *expressive, impressive* and *representational* ones. Bühler's and Halliday's understandings of function cannot be placed on the same level of object-related domains and subjective aims-in-view of their performers. Bühler's *expressive* function (*Symptom* or *Ausdrucksfunktion*) is explainable in terms of causes and effects, and his *impressive* and *semantic-representational* functions pertain to the goal-directed behavior of human agents.

In turn, Halliday's *ideational, interpersonal* and *textual* functions are determined by the aims-in-view of communicating individuals, but they are not to be equated with Bühler's functions. Correspondingly, Bühler's impressive function is carried out by perceivable means of verbal communication intended to evoke the feelings and reactions of message receivers, and a semantic or symbolic function is played when verbal means are used intentionally according to their substitutive, i.e., representational character. As to Halliday's three respective functions of language which are purpose-oriented, one has to notice that they are realized in other existence domains; being mentally concluded, the *ideational* function is oriented towards the dissemination of knowledge among individuals as members of societal groups; the *interpersonal* function, being observed, relies on initiating, sustaining and specifying the nature of the relationships between members of a particular community, and the *textual* function, being both concluded and observed is achievable by providing the structural frame for a message, and organizing the patterns of discourse relevant for a particular situation, topic or domain of interpersonal communication. Halliday's functions, the number of which has been extended in his further works, might well be compared with the illocutionary goals of speech acts, in which human individuals strive towards satisfying their needs, wants or expectations.

If a specialist in the domain of linguistics were, in turn, introduced into Fawcett's (1980: 28) cognitive view of functionalism, as related to "sociopsychological purposes of communicating minds", he might undoubtedly be aware that the so-called "functional components of grammar", as, for example, *experiential, logical relationships, negativity, interactional, affective, modality, thematic, informational, inferential, metalingual, discourse organizational functions* (cf. Fawcett 1980: 20), are modeling only "the psychological reality of language". So far, he would probably be aware that Fawcett's (1980: 23) understanding of the linguistic system as a network stored in the mind of its users that consists in a set of behavioral options, i.e., "as a code in which we have to choose between alternative types of behavior" or "as a way of displaying a number of interrelated options in linguistic behavior", is incommensurable with Bühler's (1965: 24) definition of language as a tool (existing substantially beyond the knowing subject), by means of which one person communicates to another about things, because these definitions refer to two different forms of existence of the same reality called language.

Then this adept linguists would be curious whether Fawcett was indeed entitled to mention Bühler within the same context as Halliday (1973) or Malinowski (1949 [1923]) for that matter in particular. Halliday, as indicated by Fawcett, had owed the notion of function to Bühler and not to Malinowski. But because of his understanding of language functions, which he equated with the meaning of speech acts for the satisfaction of human needs, he might be rather called the student of Malinowski, probably through the mediation of John Rupert Firth (1957 [1935]: 31) who spoke in favor of the "meaning as function in context".

> It is perhaps easier to suggest types of linguistic function than to classify situations. Such would be, for instance, the language of agreement, encourage-ment, endorsement, of disagreement and condemnation. As language is a way of dealing with people and things, a way of behaving and of making others behave, we could add many types of function – wishing, blessing, cursing, boasting, the language of

challenge and appeal, or with intent to cold-shoulder, to belittle, to annoy or hurt, even to a declaration of enmity. The use of words to inhibit hostile action, or to delay or modify it, or to conceal one's intention are very interesting and important 'meanings'. Nor must we forget the language of social flattery and love-making, of praise and blame, of propaganda and persuasion (Firth 1957 [1935]: 31).

To be recalled is that Malinowski proposed to search for the meanings of speech acts with respect to how they function while satisfying integrational, narrative, pragmatic, and magic needs of people.

What Fawcett (1980: 25) considered as an important contribution of Halliday (cf. 1973: 105–108), in comparison to Bühler was "his resignation from the idea of a dominant function" while promoting "the plurality of language functions". Moreover, as Gunter R. Kress (1976: viii–ix) highlighted, Halliday borrowed the treatment of language as a multifunctional devise from Malinowski, without adopting his proposal for distinguishing the major functions of speech, despite the fact that he made use of them in several ways. Halliday's functions, interpersonal, ideational, and textual, incorporate some of the facets pointed to in Malinowski's functions. Above all, there is quite a strong affinity between Malinowski's pragmatic and Halliday's interpersonal function.

The aim of this chapter does not consist in reviewing the various sorts of functional grammars. But, having compared the postulate of M. A. K. Halliday to include 'functional components' into the structure of the language system with, for example, Simon C. Dik's (1978) theory of the functional organization of natural languages, and Talmy Givón's (1984, 1995) understanding of functional realms, its main assumptions point, nonetheless, to the notion of communicative function related to the investigative domain of discourse pragmatics.

In fact, Dik (1987: 82) uses the term *functional* in three different, although interrelated senses: (1) taking a functional view on the nature of language, (2) attaching primary importance to functional relations at different levels within the organization of grammar; (3) expressing the desire to be applicable to the analysis of diverse aspects of language and language use.

Givón's understanding of 'function' has been specified in two contexts: the first one with regard to "functional realms" encoded in human language, as: (a) lexical semantics, (b) propositional semantics, (c) discourse pragmatics, and the second with regard to the communicative function connected only with one of the syntactic structures, namely the discourse in use. The three functional realms are, in Givón's view, hierarchized, as: meaning, information, and function.

Respectively, one can agree with Givón's (1984: 30–32) reasoning that *words* ('lexical items') have meaning but carry no information by themselves – unless they are embedded within propositions. Hence, it is possible to characterize the meaning of words without reference to either specific propositions or specific discourses in which they are embedded. Furthermore, *propositions* carry information once words are placed into them, but they do not carry any specific discourse function unless they are embedded within a larger context of discourse. However, it is possible to characterize the information carried in a proposition without reference to a particular discourse context.

Finally, one can say after Givón that, only within a specific context of discourse, propositions carry a "discursive function". Conversely, it is impossible for external observers or communicating agents to characterize "discourse function" without considering the "propositional-semantic information", just as in the same way it is impossible to characterize propositional-semantic information without taking into account the "lexical-semantic meaning" (cf. also a slightly modified model of functional realms in Givón 1995: 395–405).

4.2. Classifying linguistic functions as social domains of language use according to Charles Ferguson and William A. Stewart

Starting from the 1970s in the typology and holistic description of languages based on extra-linguistic factors, the notion of the social functions of language was in wide use (discussed in Wąsik, E. 1999d). Most of the classifications of the social functions of language, understood as place, or domain of human life where the language might be used in communicational settings, come from Charles Ferguson (1959, 1962a; 1962b; 1966a; 1966b) and William A. Stewart (1962; 1970 [1968]).[4] All functions of

---

4 The domain of language use as a constellation or set of such elements as language, communication participants and co-participants as well as the place of communication, topic (the subject of conversation or the reason for language choice were proposed to be distinguished by Samuel Fishman in his article entitled as "Who speaks what language to whom and when?" (1965: 67–88). For Ralph Fasold (1984: 67–84, mainly 77), however, as a function one should take into account is

language distinguished by both authors have been summarized by Stewart (1970 [1968]: 540), as e.g.: official, regional, colloquial (language of wider communication), international, metropolitan, group-specific, educational (medium of instruction), scholar (subject of study and school subject), literary, religious.

Abbreviating Fergusons and Stewarts classificatory proposals to three selected social functions, one can state, after Ralph Fasold (1984: 77 and 1989: 113–115) and Ulrich Ammon 1987: 230–263) conditions, according to which a given language is called: (1) an official language when it serves by the power of legal institutionalization and standardization as a means of: (i) spoken communication among government functionaries performing their duties; (ii) written communication between and within government agencies; (iii) record keeping and the publication of legal documents and regulations; (2) a national language, when it serves as (i) the symbol of identity and aspirations for a sizable powerful part of the population, while being (ii) widely used in everyday communication, (iii) spoken fluently in a country or state, (iv) acceptable as an authentic common property, related to a perceived glorious past; (3) a regional, provincial, metropolitan, or a group-specific language, when it is used: (i) as a means of everyday conversations by all members of a given speech community in a given region, province, capital city, profession, or interest sphere, (ii) as an integrating and separating factor, etc. (cf. Ammon 1987: 256).

The term *social functions of language* reflects its understanding in sociology (cf. Hempel 1965 [1959]), under the influence of the philosophy of biology where it was derived from the search for the natural functions of a given element, respectively an organ for the sake of a given goal-directed organism; to detect such a function meant in effect to ask how a given organ contributed to the goals of an organism as a whole. In sociology the meaning of the term *function* was identified with the consequences of how a given element acts being driven by the requirements of a given society.

The departure from the term *function* activated an unrestrained listing of tasks a given element might fulfill, and ended in explorations for all possible equivalents of a given element in the society (for a detailed discussion see, *inter alia,* Ammon 1987: 235). As a consequence, the accep-

---

only that which is practiced, and not as what is imposed by law which is not applied.

tance of the term *function* from sociology to the domain of the neighboring disciplines of linguistics resulted in the belief that scientists are able to discover an undetermined and immeasurable number of the so-called functions of language. It seems, however, that the division of social functions practically connected with the detachment of the domains of life, and situations of language use, can be also identified with the search for aims or tasks, around which communicating individuals unite themselves into social linkages.

5. Against the indiscriminate enumeration of linguistic functions

To sum up, one should say that different faces of linguistic functionalism are conditioned by the application of incommensurable frames of reference to different modes of existence and manifestation forms of language as a holistic phenomenon or as a system of lexical and grammatical elements in its textual realization. Ultimately, it has to be stressed out that searching for linguistic functions theoreticians of language sciences should be aware that the foundation of linguistic functionalism results from different views on language as a natural phenomenon and cultural artifact, and that the material object of linguistic studies may be viewed from different disciplinary points of view. Instead of the functions of language as a means of communication, one should rather distinguish the functions of the manifestation forms of language subsumed under other kinds of objects.

Those who indiscriminately search for the functions of language have to bear in mind that the number and the kinds of functions result from the differences in the definitions of language itself and from the fact that the material object of linguistic studies, due to its heteronomous nature, may be also studied from other non-linguistic perspectives. Therefore, taking a holistic approximation to language as a natural and/or cultural object, it is important to find the sources of the various frameworks that have shaped so many of the faces of linguistic functionalism.

On specifying the habitat of the functional sphere in language, the linguist has to consider both its concrete and mental forms of manifestation. Dealing with the functions of language as a system in abstraction from the environment of its speakers, linguists (cf. Wawrzyniak 1974 and Wąsik, Z. 2006a: 45–47) have highlighted at this point as many as two, four or six modes of its existence, as: (1) in concrete meaning bearers, (2) in concrete activities of sending and receiving the meaning bearers, and (3) in relational values of concrete meaning bearers; as well as (4) in mental equivalents of meaning bearers, (5) in mental activities of conceptualizing and interpreting the meaning bearers, and (6) in associations of relational values of meaning bearers memorized by communicating individuals.

From a human-centered perspective one has to remember that language exists also in communicating activities of people who send and receive messages while referring them to the commonly known extra-linguistic reality. The results of the study have shown that language also manifests itself in two other additional existence modes, which are in fact determined by its communicational nature.

Thus, language as a social activity exists also (7) in concretely observable, dynamic interactions between people, that is, in interpersonal linkages of physical nature, when communicating individuals are bound by sound waves and other surrogate codes, transmitted and received in communication channels, and (8) in logically concluded intersubjective linkages, based on the assumption that people interpret meaning bearers in a similar way while referring them to the commonly known extra-linguistic reality, that is, in linkages between the minds of individual communication participants. Hence, the inquiries into language as a tool should end with the conclusion that the main function of language consists in establishing linguistic communities of meaning cognizers and meaning knowers, as well as meaning producers and meaning interpreters among communication participants on various levels of social groupings with reference to the realization of purposes and tasks of people, whom it serves.

It seems evident that other kinds of functions may be distinguished in language understood as a system of communicative means, and others also in the language treated as an instrument of mental cognition, and still others – in the context of receptive and productive activities or the communicative faculties and habits of language users. Thus, a linguist must be aware of the fact that his object, language as a system, is situated within other systems, i.e., becomes a member of the same paradigm in other kinds of settings. In such cases, we do not describe the function of language as a means of communication but rather the function of language as a kind of other object, e.g., of art, of ritual, of behavior, of national symbol, of entertainment, as a social institution.

Without specifying, whether one has in mind the natural language as an exponent of mankind, ethnicity or group identity, language varieties, standards,[5] dialects, professional jargons, speech acts or speech genres of communication, linguistic faculties and cognitive abilities of humans, or constituents and structures of verbal means realized in various contexts of language use, etc., we might encounter in linguistic works a number of the so-called language functions, which reflect in fact the satisfaction of individual needs or the social requirements of people through the realization of communicational tasks, such as, in an alphabetical order, for example, argumentative,[6] cognitive, collaborative, communicative, competitive, conflictive, controlling, convivial, deceptive (to make perceptually misleading), deferential, degrading, delimitative, developmental function of language abilities (connected with the child language), diacritic, discriminating, distortive (to give a false, perverted or disproportionate meaning), emotional, enculturating, evocative, excluding, experiential, expressive, heuristic (learning generalizations about someone's environment), ideational, identifying, imaginative (creating one's own world), impressive, conative, including, informational, informative, instrumental (satisfying someone's needs by the use of protolinguistic symbols), interactional (mediating someone's togetherness with people), interpersonal, interpersonal, logical, ludic (realizing entertainment purposes), manipulative, matetic "macro-functions" (personal, heuristic, imaginative; language is used to explore the environment, i.e., language as learning), metafunctions of language (reflected in the internal organization of language, for example, in the systems of Transitivity, Mood, and Theme), metalingual, persuasive, phatic, pragmatic "macro-functions" (instrumental, regulatory, interactional; language is used to manipulate the environment), pragmatic, prevaricating (to speak falsely, misleadingly, or so as to avoid the truth; deliberately misstate; equivocate; lie), regulatory (using language to order the adults about), representative, ritual, semantic,

---

5   The function of standard variety of language sanctioned by social pressure was discussed by Paul Lucian Garvin and Madeleine Mathiot (1970 [1968]).
6   As Geoffrey Neil Leech (1990: 48–58) points out it is the use of language for reasoning and logical operations within a mind according to Karl Popper (1972: 70, 106, 117).

separating, significative, stimulative, symbolizing, textual, transactional, unifying, etc.

To ponder the appropriateness of the uses of the term *function* to language, as, e.g., to describe the ludic, deceptive, or therapeutic role of language, we would propose to consider the following example based on the philosophical reasoning of Andrew Woodfield (1976) and Larry Wright (1976): Imagine that you have jammed a newspaper under the door so that the door, in consequence, does not close. Can you say that it is the function of the newspaper to block the door? The answer should be: "No, of course not." The newspaper has in reality another basic function, it has to bring news collected and edited by journalists. True is only the fact that the respective newspaper, as a type of object, may be subsumed under another type of object, namely, in this particular case, an obstacle. A newspaper may be considered as being sufficiently good also for other purposes. But in such a case it does not exist anymore as a newspaper serving its primary function. As a result we can say that the use of the verbal means of communication is good to entertain, to tell lies, or to console someone. But there are also some other nonverbal means that can be used for the same purposes.

Chapter Three

The linguistic properties of communicating individuals and their role in the construction of an intersubjective world of meanings

The subject matter of our discussion constitutes those properties of language that unite people as communication participants in their natural and socio-cultural environments into interpersonal linkage systems based on the realization of common communicational tasks. It is assumed that the basic function of language is to establish homogeneous communities of meaning cognizers and meaning knowers, as well as meaning producers and meaning interpreters among communication participants. Language can be regarded, therefore, as a relational property of man, because it may be deduced, firstly, from the observed communicative interactions that come into being, and secondly, from concluded correspondences of comprehension that can be attained.

1. Language as a creator of communicative collectivities

Thus, one can state, that people are organized into collectivities (communities) of social, national and international character on the basis of both physical and psychical relationships. In the first case, communicating individuals are linked by sound waves playing the role of verbal messages as meaning bearers, as well as their surrogates being in turn transformed again into sound waves in the process of their reception, which are accessible to empirical observation and experiment. That is to say, when people are engaged in the activities of sending and receiving messages, the existence of linkage systems is connected with the expenditure of certain amounts of energy that is measurable. In the second case, communicating individuals, as participants of group communication, enter into the intersubjective relationships when they cognize and interpret the meaning of verbal messages in the same or similar way in accordance with their referential value. These relationships are dynamic in nature because they find their reflection in the mind of communication participants in the form of changeable linguistic knowledge about the concluded reality.

2. The linguistic properties of man against the background of constructivists' theories

The frame of our reference will adopt a constructivists' perspective based upon the positions held by Victor Huse Yngve who promoted a hard-science view of human linguistics (the concepts of which he developed in his subsequent publications beginning in the 1970s) and Franciszek Grucza who considered language as a property of an individual speaker/hearer forming herself/himself in communicational activities. As has been revealed by Krzysztof Korżyk (1999: 14, cf. especially footnote 5), certain parallels between the positions of F. Grucza represented in his book on metalinguistics (1983) and V. H. Yngve summarized in his book *From Grammar to Science* (1996) may be stated, which highlight the fact that both authors stress that the primary object of linguistic studies are people. However, while Yngve proposes to detach the "linguistics of people" from the "linguistics of language" thereby devoting more attention to the creation of linguistic communities along with their environments as steadily growing, long-lasting and temporary-disappearing interpersonal linkages, Grucza by contrast exposes, within the framework of metalinguistics, the relevance of linguistic-communicational capabilities and the aptitudes of speakers/hearers, defined in terms of structural systemic and functional linguistics (cf. the discussion of Grucza's anthropocentric conception of language for the tasks of applied linguistics by Z. Wąsik, 1986).

An examination of Yngve's and Grucza's presentations of the linguistic properties of man against the background of constructivists' theories will be advantageous for two reasons. It will allow us to cross beyond the boundaries of linguistics proper and to understand how communicating individuals, as participants of group communication, contribute to the creation of common knowledge instituted in the language of social collectivities, in the same way as scientists put forward their theories modeling the extra-linguistic reality which is intersubjectively comprehensible. At this point, it would be appropriate to make a reference to Piotr Landsberg (2003: 135) who stated, discussing Samuel Huntington's (1996) concept of multiple civilizational worlds in the light of social constructivism, that scientific knowledge is the product of researchers determined by the social conditionings of investigated reality. From such a point of view, considering the perspective of the contemporary philosophy of science, the value of social conditionings of science results from the social environment, in which scientific research is conducted, and not from the relationships of scientific theories to their objects of reference, which does not contradict the fact that scientists strive to present and model physical reality as truthfully as possible (cf. Yngve 1996: 121).

3. On the personal constructs and social construction of reality

Constructivism as an investigative perspective, which is the subject of our interest in relation to man, has assumed two positions, first, as the personal construct theory in psychology, and second, the social constructivism in sociology (cf. Wąsik, E. 2006: 87–1001) but according to its applicative distribution in science and education it started to turn from a general moderate investigative standpoint to that of a radical extremist attitude.

The term *personal construct* was coined by George Alexander Kelly in 1955. Following the distinctions made in Kelly's book *The Psychology of Personal Constructs*, this term refers to the cognitive manifestation of the social, physical and psychical world in the mind of an individual human being. On the assumption that the process of cognition is based on the interpretation of cognized phenomena and the ascription of meanings to them in steadily changing contexts and situations, such a personal construct is said to be developed in the mind of a cognizing individual. The mere notion of personal construct, understood as a cognitive construct or the scheme of the interpretation of reality, suggests that every individual, as a cognizing subject, subsumes concrete objects to certain classes of objects with regard to their referential meaning. In the opinion of the followers of constructivists' theories in psychology, discussed by Zbigniew Nęcki in his book on interpersonal communication (2000: 31–33), and Teresa Siek-Piskozub in her paper on "Social constructivism in foreign language education" (2004: 11ff), social communication can lead to creating similar personal constructs in the minds of people interacting within the same culture. These constructs form the basis for a similar perception of the world as well as unified behaviors against the objects evaluated with respect to their utility. Such personal constructs, expressing subjectively defined referential meanings, constitute the most important factors which determine all forms of social behavior, including the verbal form of communication. When similar conceptual constructs come into being in the minds of members of a certain linguistic community, as a result of recurring interactions, it is understandable that they find reflections in the commonalities of the expressions of meaning bearers. People integrate with each other individually on the basis of observable verbal means of expression and in accordance with inferable comprehension of meaning bearers.

The fact that in human communication the exchange refers not only to material objects but also to ideas and meanings was exposed also by such sociologists, as Piotr Sztompka (2002: 77–78), *inter alia,* points out while citing the founders of Symbolic Interactionism, Charles Horton Cooley and George Herbert Mead. One can even say that linguistic communication realizes itself through such interactional and transactional behavior of people within and/or with their environment that contributes, in effect, to the diffusion of knowledge resulting in the creation of homogeneous communities of interpreters who possess the same knowledge about extra-linguistic reality.

Having been elaborated upon in the domain of sociology, the theory of the social construction of reality is based on philosophical theories dealing with the relationships between the inner and outer world of man. It expands however beyond the theory of cognition and explains the way in which human individuals accumulate their knowledge about the outer world, the investigative questions of which are focused around the theory of perception.

The foundation of sociological constructivism constitutes an assumption that social reality is shaped by information gained by particular human beings as organisms in interaction with their environment. As pointed out by Peter Ludwig Berger and Thomas Luckmann, the authors of *The Social Construction of Reality*, originally published in 1966, man is a social being and his contacts with external environments is mediated by symbols. In the opinion of Berger and Luckmann knowledge related in a certain way to reality, is incessantly connected with certain contexts and social situation, insofar as it is always created by society and transmitted among its actual members. However, apart from social factors, this knowledge is determined in nature by historical, psychological as well as biological factors. Nevertheless, following the distinctions of the French sociologist Émile Durkheim, Berger and Luckmann (cf. 1966: 60 and 67) have explained that their views do not contradict the conception of the German sociologist Max Weber "of the meaningful character of a society. Since [as they argued] social reality always originates in meaningful human actions, it continues to carry meaning even if it is opaque to the individual at given time" (1966: 197, endnote 27, addition in brackets is ours: EW). This is especially true when one states that society consists (as a sphere of objective facts), first of all, from the externalized products of human activity, in which subjective meanings are personified. In other words, objective facts, which constitute social actions, are equipped with subjective meanings and it is: language which "objectivates the shared experiences and makes them available to all within linguistic community, thus becoming both the basis and the instrument of the collective stock of knowledge" (Berger and Luckmann 1966: 68).

Following the conviction of social constructivists, society is the creator of knowledge, although an individual human being as an organism experiences, de facto, reality while receiving various kinds of information from the environment. The stock of everyday knowledge is created due to social interactions; this knowledge is – as one can say after Berger and Luckmann – negotiated and approved among particular members of society (cf. 1966: 19–46). A certain kind of a social construct is the reality of everyday life or the world of life, which comes into being as a result of communicational activities. The reality of everyday life or the world of life is considered as one of many realities, albeit a basic one. But it is not identical with the really-existing objective world. As a result of interactions it becomes an intersubjective world, that is, the world which is shared by an individual with other individuals. As Berger, and Luckmann state:

> The reality of everyday life further presents itself to me as an intersubjective world, a world that I share with others. This intersubjectivity sharply differentiates everyday life from other realities of which I am conscious. I am alone in the world of everyday life without continually interacting and communicating with others as it is to myself. Indeed, I cannot exist in everyday life without continually interacting and communicating with others (1966: 23).

To be noted at this point is that the term *world of life* was used by the German sociologist Jürgen Habermas in his book *Theorie des kommunikativen Handelns* (1987 [1981]) as a correlate of moving closer to comprehension. He believed that acting subjects reach a state of comprehension always on the basis of the world of life. As Habermas claimed, members of communicative communities consider the world of life as non-problematic, fairly objective, while detaching it from the intersubjectively shared social life and the intersubjective world of communication. Society construes this reality of everyday life, and thanks to this an individual lives within a common world with people, sharing with them common knowledge. Apart from this, one can distinguish in everyday life such spheres which become habitual for a given person or some other spheres, the understanding of which is problematic and in which it would be difficult to function. Thanks to social interactions in everyday life, that is, in immediate contacts – in face-to-face interactions an individual has access to the subjectivity of the other.

As Berger and Luckmann argue:

> Human expressivity is capable of objectivation, that is, it manifests itself in products of human activity that are available both to their producers and to other men as element of a common world. Such objectivations serve as more or less enduring indices of the subjective process of their producers, allowing their availability to extend beyond the face to-face situation in which they can be directly apprehended (1966: 34).

In their view, human expressivity manifests in products being accessible both to their creators and to other people. These real objects, which are observable and which become symptoms of actions – or their meaning bearers, Berger and Luckmann consider as elements of a common world. Noteworthy, among such elements are, e.g., bodily symptoms, gestures, postures, certain movements of hands legs, etc. It is these which are accessible to communicating individuals in immediate contacts. In the case of verbal understanding – communication by voice – where sound waves being produced are recognized as elements of common world, the objectivation – in the view of Berger and Luckmann – comes also into being, as we may conclude from the following statements:

> A special but crucially important case of objectivation is signification, that is, the human production of signs. A sign may be distinguished from other objectivations by its explicit intention to serve as an index of subjective meanings. To be sure, all

objectivations are susceptible of utilization as signs, even though they were not originally produced with this intention (Berger and Luckmann 1966: 35).

To sum up, the achievements of constructivism, as a set of theories, continued by Ernst von Glasersfeld (1995, 1988, and 2001), and conceptually applied, *inter alia,* by James Moffett (1983, 1987), Margaret D. Roblyer, Jack Edwards, Mary Anne Havriluk (1997), and which have in consequence further developed into an extreme formal approach by Alexander Riegler and others (2001), may be summarized in at least three conclusions (cf. Wąsik, Z. 2006b: 22–24).

Firstly, constructivism as an investigative perspective is founded upon the assumption that people create their own view of the world they live in on the basis of reflections of their individual experiences. Hence, constructivists expose the role of an individual self as the cognizing subject and maker of meanings. Each individual is regarded as generating his or her own mental model which allows them to understand or to make sense of the world by selecting and transforming information, formulating hypotheses, and coming to decisions that rely on his or her personal cognitive structures.

Cognitive structures provide the basis for meaning creation and deciphering through mental schemata or models organizing the experience of an individual, which allow going beyond the information provided to him or her by sent and received meaning bearers. Considered in the educational context, discovering, learning and communicating are specified in constructivists' theories as the searches for meaning. In such processes, individuals, viewed as lifelong learners, discoverers and/or communicators, constantly modify their personal mental models to accommodate to new experiences by construing new concepts or imagining new ideas with reference to the knowledge which they already possess.

Secondly, while cognitive constructivists treat the individual as a personal scientist who creates and understands the meaning of phenomena organized in his or her mental world, social constructivists by contrast ask how personal meanings of communication participants enable him to understand the meaning bearers arising out of group interactions, defined in sociality terms as meaning-negotiating activities. One of the examples of social constructivism constitutes the so-called sociology of knowledge in the domain of science and education which claims that the personal constructs of communicating individuals are distributed in the form of reported speech and may contribute to common understanding processes only as types of shared experiences. This is based on the assumption that when one communicator employs a mental construction reflecting his own experience, which is similar to that employed by another, then they both may understand each other effectively; supposing that his or her construction processes appear as mentally similar to those of the other communicator, then he or she may play a role involving that particular individual in the social process as communication participant.

Thirdly, the constructivists' position to the personal property of meaning results from the social character of language and human communication, hence: (a) meaning being a human construct is dependent upon the person who makes it; (b) meaning does not reside in the words, symbols or appeal signals with which individuals express their emotional and conceptual contents, and, as such, (c) meaning cannot be passed on as an entity in the same manner as meaning bearers; thus, (d) language has to be regarded as a behavioral system of meaning bearers which trigger communicating activities within the cognitive domains of particular communicating individuals; and (e) communicating in a given language is based on the continuous mental processing and interpreting of meaning bearers, which are being produced and received; whereas mutually shared meanings happen only to be assumed.

More profoundly rooted is the concept of "radical constructivism", developed in the context of science which considers any kind of knowledge as constructed rather than perceived through the senses. Radical constructivists (Glasersfeld 1995, 2001; Riegler 2001) claim that the cognizing subject has no other alternative than to construe what he or she knows relying on his or her own personal insight. This is so because all kinds of experience including interpretation of language are essentially subjective. For radical constructivists it is useless to think about scientific knowledge as representing external reality independently of a knowing subject. Similarly, one cannot accept the belief about the existence of a universal language, which might represent external reality in terms of so-called objective knowledge. Radical constructivists do not reject the existence of the world populated by mind independent items, but science for them is a form of cognitive activity of researchers who strive towards a subjective organization of their experiential domain and not to the discovery of an objective world that exists beyond their capacities of cognition. For the representatives of radical movement in education, experiences in particular must fit into the current web of knowledge in general thus to make sense for the development of an individual scientist. Seen as a mind-depended entity, the scientist appears as an epistemological solipsist being unable to transcend the domain of his individual experience. The scientific theories he encounters or construes on his way to apprehending knowledge as sets of propositional contents appear only as conceptual models that help him only to manage his experiential domain. Models may be in the course of time replaced by others based on concluding, slightly altered or entirely innovative, constructs when the experiential domain of the scientist expands as a result of his subsequent research activities (cf. Glasersfeld 1988: 83).

4. Man as the object of linguistic studies in Franciszek Grucza's and Victor H. Yngve's depictions

Having discussed the foundations of social constructivism, we will now present the conceptions of two authors who have affirmed that they are interested in people from the viewpoint of their linguistic properties abstracted from their social environments. To begin with Grucza's views (cf. 1983: 416–418), man possesses such communicative properties, understood as his aptitudes or abilities, which enable the communication participants to attain mutual understanding with each other by means of linguistic utterances. In other words, these properties allow them to create, distinguish, delimit, and recognize semantic signals, while interpreting and appreciating their communicative values. Aptitudes or abilities in the linguistic-communicational domain develop and form themselves due to participation in communicative events by means of speech and writing. Among the linguistic properties of speakers and/or hearers Grucza (1983: 422–428) enumerated, such abilities, the possession of which can condition effective verbal communication, as, for example, the ability to formulate a report, a protocol, to write a leaflet, an advertising slogan, an article, an essay, a review, or the ability to deliver an occasional speech, to express congratulations, propose a toast, to introduce speakers or topics of the meeting, etc. Moreover, in turn, Grucza proposed to consider the development of such communicational-linguistic properties of man that constitute, depending on the situation, the abilities of selecting the appropriate means for eliciting, conducting and maintaining conversation, discussion or debate, or for the purposes of convincing someone about something, for persuading someone into or from something, or to mitigate or moderate conflict. Grucza classified also the abilities among the properties of humans expressed in language rules that enable the communicators to cooperate effectively. Furthermore, he saw also the need to distinguish form-related properties, such as substantial and grammatical properties, or function-related properties, such as semantic, semasiological, onomasiological and pragmatic properties (contextualizing and con-situational properties, while exceeding the boundaries of language, as, for example, stylistic properties, rhetorical, dialectical and hermeneutic). With regard to the systemic-structural properties, he proposed to distinguish morphemic properties, lexemic, phrasemic, and other properties. Apart from the language-related properties Grucza saw also the need to enumerate various functional properties of linguistic communication, as expressive, representational, impressive, emotive, referential, poetic, phatic, metalinguistic, conative or furthermore, interrogative, exclamative, imperative, directive, and adhorative in one dimension, and anagnostic, diagnostic and prognostic in another terminological dimension. And, among the latter also: descriptive, explicative, comparative argumentative properties of language, or in more precise terms, in Grucza's belief, the properties of speakers/hearers (cf. also Wąsik, Z. 1986: 96–98). In addition to this classification, which has shown a survey of linguistic functions and categories, one should underline that the understanding of "meaning" has been referred, in Grucza's conception, to the properties of concrete utterances in verbal communication ascribed to utterances by their speakers/hearers.

In Victor H. Yngve' view, the linguistic properties of a communicating individual taking part, as a real person, in group interactions constitute the core of the theory of human linguistics considered from a hard-science perspective. To describe the individual being active communicatively in a concrete situation where more than two persons interact, Yngve proposed a narrower theoretical notion, in which he referred to the properties of a participant of communication. In Yngve's view participants of group interactions execute certain activities which are directed to the realization of communicational tasks, such as, for example, agreeing and disagreeing, buying and selling, teaching and learning, negotiating or arguing and convincing, haggling over price, etc. (cf. Yngve 1996: 84, 86). As linguistic properties, Yngve (1996: 123–130) proposed to distinguish such communicatively relevant properties of a person that can be described, as, for example, that someone is bilingual, or that he or she possesses a specific accent or knows how to address others, knows how to pose questions or how to interpret utterances depending upon an actual context. Yngve saw also the need of describing persons as group members and their behavior and other linguistically relevant constituents and surroundings in terms of real objects, which are physically observed on the level of concrete data, called an assemblage.

However, on the level of theoretical constructs communicating individuals as group members become participants of the so-called linkage. Some other constituents of a linkage form, apart from participants, channels, props and settings. And so, for the theoretical framework of Yngve's human linguistics: (1) channel, is a representation of energy and means of energy flow that are important for communicating activities that involve more than two person, i.e., a linguistically relevant sound, an associated energy of vocal sound waves, or a flow of light energy associated with gestures; (2) props are material representations of real objects related to anything that could be interpreted as relevant for signs, tools, and other immanent phenomena underlying physical tests; (3) settings are representations of elements of the physical surrounding of a group in an assemblage that are required for the understanding of communicational contexts. In groups, in which interpersonal communication takes place, one can observe specific phenomena of a linguistic character, because every group is connected with a specific way of interaction, with specific external conditionings or circumstances or specific relationships between observed phenomena. Yngve expressed his conviction (cf. 1996: 19) that there are certain linguistic properties of persons that come into being only in the case when they act as members of a group, for example, agreeing is possible among participants communicating at least in a couple. As regards the properties of a real person, its boundaries approximately agree with the scope of a biological organism. Nonetheless, in the case of a communicating individual as a participant in a linguistic linkage, the researcher can arbitrarily establish, as Yngve (1996: 176–177) points out, the limits and the scope of the domain of his studies.

For the purposes of describing group communication Yngve (1996: 176–180) postulated to distinguish various kinds of linkages, in which members of small groups communicate in face-to-face interactions, while large groups embrace an immense number of participants, that is, where the boundaries of communicative communities are not known and the groups, which are of medium size, where the number of its participants can be measured, as, for example, professional organizations, schools, meetings, etc.

Another dimension of group divisions was proposed with respect to linkages which might be open or compact. In such a context, Yngve characterized different kinds of interrelationships between linguistic linkages, while at the same time noticing that some constitutive elements of linkages overlap and some do not when one linkage covers another linkage or two or more linkages at the same time. In linguistic linkages people form a certain network of hierarchical relationships.

Similarly, as in the case of individuals being linked in group interactions, one can say that there is an interaction between separate assemblages of group members and their linguistically relevant exchanges of energy, objects and surroundings are coupled immediately or through linkage of a third kind. Contacts between coupled linkages occur, when particular groups of interacting people are united also by such communicational constituents as channels, props and settings. It happens also, as we may conclude further from the conception of human linguistics, that linkages can be coupled also by their internal arrangement, especially in the case of larger organizations. In fact, these distinguished dependencies between communicating individuals and groups constitute objects of the real world that can be tested in empirical observation and experiment.

One should stress that the linguistic linkages as communicative communities develop and last due to interactions between its participants. The occurrence of interaction can be noticed in the so-called physical domain, when sound waves and other means of energy are observed as linguistically relevant objects linking participants of communication with each other. Thus, the interpretation of phonic sequences, that is the ascription of meaning to them in terms of referential linguistics, based on semiotic-grammatical tradition, either by communication participants or by researchers can be the subject of inferences, which is made, while alluding to Yngve's distinctions, exclusively in the so-called logical domain.

Being interested in man as a person and interpersonal relationships, and investigating the functional reasons of linguistic linkages for which they are usually created, Yngve (1996: 186) introduced the explanatory term, namely, the communicational *task*. In this way he wanted to study the activities of persons and groups in linguistic linkages in terms of behaviors directed towards the realization of certain hierarchies of tasks and subtasks which are performed and executed.

Despite the assumed hard-science approach to human communication, Yngve could not manage to remain in the physical domain while making inquiries into the linguistic properties of people from the viewpoint of physics, chemistry and biology, as he claimed. This is especially evident with respect to the distinction of tasks, which the participants of group communication are said to fulfill.

What kind of tasks can be ascribed to communicating individuals in particular or collectivities in general depends only on the subjective inferences made both by communication participants and researchers who act in any case as receivers. Enumerating the tasks, which the participants of communication should fulfill, is nothing less that entering the logical domain of subjective inferences, in which an observer, somehow from the outside, will ascribe to communicating individuals and collectivities certain determined tasks.

The same thing happens when the addressees, as receivers in communicational acts, are able to interpret the referential values of meaning bearers entirely on the basis of introspection. In other words, meanings or tasks, in the same way as values in the context of axiological investigations, are to be ascribed to phonic sequences or to other physical objects, when they are relevant from a linguistic point of view. Hence, Yngve's theory can be appreciated as useful in search of the linguistic-communicational properties of man but only on the level of facts, which are observable.

Meaning bearers, both in verbal and non-verbal interactions, as, for example, objects that are linguistically relevant, and, in the case of language sound waves, being sent and received as linguistic utterances, underlie always the interpretation of receivers from a solipsistic outlook. In the real world, only the communicational means constitute the objectively available components which give one or more persons the possibility to access the subjectivity of another person. One can say, therefore, appealing to the tenets of constructivists, that they are deciphered on the basis of typization of the firstly cognized and then subsequently known phenomena.

5. The accessibility of the linguistic properties of communicating individuals

The departure point of this paper constituted the thesis that language does not exist as an autonomous phenomenon. On the contrary, it must be investigated, with due reference to its origin and means and ways of functioning, exclusively in relation to the communicating individuals who are accessible to external observers as an empirical reality. When people communicate it is perceptible in the first instance that extraorganismic products of individuals in the form of sound waves and energy flow, but how people refer and interpret them, as information carriers, is a matter of assumption. Observable also are the changes in communicating individuals when they interact with other individuals, so that they can form interpersonal linkages.

Having considered the viewpoints of anthropology, psychology and sociology, in the first methodological dimension, as well as biology, philosophy and culturology in the second, we postulate to accept in a human-centered approach, the following investigative perspectives, which appear to be useful as a set of postulates for defining the scope of heteronomous linguistic studies, as: (1) collective solipsism, (2) communicational pragmatism, (3) social constructivism, (4) "enactive" ecologism, (5) experiential "recentivism".

(1) Representatives following the attitude of collective solipsism assume that language constitutes the common property of individual "selves" uniting them into the members of society, which form a set of collectivities and not collections of individuals. The advantage of solipsism – that recognizes the special role of a cognizing and knowing subject, as an organism, being aware of his/her/its ego-quality, on the basis of an individual reception of experiences – is especially emphasized in the work of the German biologist and founder of modern ethology, Jakob Johannes von Uexküll, who searched for the patterns of animal behavior that occur in their natural environment. In Uexküll's philosophy of biology (popularized in Poland by Aldona Pobojewska, by her thesis on the biological *a priori* of man, 1996, and the selection of Uexküll's writings, cf. 1998: 13ff) human organisms usually function in the role of a subject constituting the central point of his own *Umwelt*, and not as an external observer. From a solipsistic perspective for particular human beings it is impossible to find similar subjective universes, because the external

world is not identical when it is governed by the law of need-oriented causality. Collectivism is taken into consideration when the concept of the self is introduced into the sociological understanding of knowledge. Only due to the integration of human individuals, that is, when communicative actions are directed towards the self as a member of a collectivity, the group may exist as bound by permanent links. Thus, the group becomes a compact system of concatenated items in the sense of a linguistic community, when it makes use of the same verbal means of communication for satisfying the same communicational needs and social requirements.

(2) Those practitioners who depart from the viewpoint of communicational pragmatism, represented, *inter alia,* by Geoffrey Neil Leech (1983) especially in his *Principles of Pragmatics*, believe that language should be studied in relation to interpersonal communication as a set of activities forming temporary linguistic linkages and long-lasting communities through the realization of the communicational tasks (intentions or goals) of individuals in respective speech acts. Distinguishing the two domains – the physical domain and the logical domain within the investigative field of pragmatics of linguistic communication – one should state that communication takes place when people talk and understand each other while producing verbal means of similar referential value. The occurrence of mutual understanding, however, which is based on concluded reality, belongs, as a mental fact, to the logical domain. Considering the constituents of a concrete speech act, the physical domain unites, firstly, the people who communicate, and secondly, the physical sound waves as phenomena uniting the speaker(s) and the hearer(s), and thirdly, all those physical objects and other elements of the surroundings that are relevant for the realization of the tasks of communication participants. In other words, one could say that only these three elements: communication participants, material bearers of human intent, and the situational context, constitute what can be empirically ascertained and proved, i.e., what forms a concrete observable whole. Nothing can be said with certainty, without logical reasoning, about the meaning or force of utterances, about the referential value of what the people mean, whether their communication is successful or not.

(3) Social constructivists claim that communicative communities form not only common meaning bearers but also the intersubjective world of meanings based on inferential knowledge about the reality, to

which they refer those meaning bearers for achieving a mutual understanding. As pointed out by Siek-Piskozub (2004: 11ff), with regard to the psychosocial aspects of glottodidactics, while making reference to *The Psychology of Personal Constructs* by George A. Kelly (1955), constructivists are of the opinion that knowledge is created and not discovered by the mind of humans; hence objective knowledge cannot be said to exist in reality.

(4) Proponents of "enactive" ecologism maintain that what people are talking about must first exist in the lived histories of their sensorial-corporeal interactions with their natural and socio-cultural environment, and then appear in the intellectual contents of their communication; In the theory of embodied semantics, it is claimed that man as a biological organism cognizes reality only through stimuli acting on his senses. An „enactive" concept of meaning has been developed in *Anthropological Linguistics* by William A. Foley (1997) for the tasks of linguistic anthropology, while making reference to the studies on meaning in the realm of living organisms conducted by the Mexican biologists, Humberto R., Maturana, and Francisco J. Varela (1987) in *The Tree of Knowledge: The Biological Roots of Understanding*, which were later continued by Francisco J., Varela, Evan Thomson and Eleanor Rosch (1991) in *The Embodied Mind: Cognitive Science and Human Experience* (cf. also *Linguistic Anthropology* by Allessandro Duranti 2000 [1997]). In this context, one should also duly note the concept of "subjective significance" as discussed by Zdzisław Wąsik in *Epistemological Perspectives on Linguistic Semiotics* (2003: 107ff).

(5) Following the conception of experiential "recentivism", it is assumed that for an individual human being communicatively relevant with regard to his actual and prospective acts of comprehension and interpretation are solely his previous experiences with an outer and inner world. In this particular context, due reference should be afforded to the theory of Józef Bańka (1986) pertaining to the ontology of actual being.

How to access the knowledge of individuals about the real world in the form of mental reflections of notions attached to certain linguistic expressions and utterances and, in consequence, how social meanings are preserved in a given language, on the basis of which people form communicative communities, are crucial investigative questions posed by practitioners of cognitive linguistics. The idea of linguistic comprehension is illustrated by the imagination of an intersubjective world, advocated by Leech (1990 [1983]: 51–55), that is, the world, which exists due to the fact that people as subjects of interaction establish and confirm meanings of linguistic units in verbal communication. The notion of intersubjectivity, relevant especially for explaining the essence of linguistic comprehension, has been considered also by Allessandro Duranti (2000 [1997]: 255). From the assumptions regarding the investigative perspective of „enactionism", it may be concluded that the experience and biological features inherited by people, are never similar, as far as it is impossible to state the existence of the same domain of reference common for all communicatively linked people, that is when individuals participating in different communicative linkages do not create or understand the same or similar meaning of linguistic units, expressions or utterances. Hence, one could agree with the claim, signaled above, that objective meaning does not exist at all since it is embedded in the lived histories of individuals as organisms. Nevertheless, a similar knowledge of the meaning can be assumed, when the subjects of interactions happen to possess similar experiences in the histories of their life. Such an understanding of meaning, reducing its manifestation to the subjective universe of an individual, explains why there is a necessity to negotiate the meaning of verbal expressions and utterances by the participants of communication each time during conversation. It is the reason why the verbal communication of man is investigated with regard to axiological aspects of communicational purposes, as as far as its objects reference constitute ideology, values, ethics, and the various subjective needs realized by verbal means in spoken and written manifestation forms of language.

Chapter Four

On the ecosemiotic existence mode of language in local and global linkage aggregations

In this chapter, the ecology-related terms and the ecological way of reasoning encountered in the works of the practitioners of language sciences will be discussed from a historiographical and methodological perspective. Its point of departure constitutes the metaphorical term *ecology*, initially connoting the study of the relationships between organisms and their environments, which has been originally popularized in the domain of social studies concerned with the spacing and interdependence of people and institutions, and then introduced into the domain of linguistic studies dealing with external conditionings of natural languages.

In the first instance, it will be shown how the ecological-relational properties were opposed by linguists to lexical-relational, lexical-inherent and grammatical-inherent properties of language as a system. In such a view, the system of a language was seen as an ecological specimen embedded within a more complex system at a higher level, the so-called language ecosystem, which included grammar and lexicon and the people who communicate in natural and cultural settings with their dispositional and behavioral properties. The attention of linguists was then paid to observable changes within a given ecosystem in search of environmental factors that influenced the variability of language structures and elements.

Bearing in mind that the basic parts and aspects of the so-called "ecology of language" comprise the external and internal conditionings of communicating individuals and communities, substantial arguments, in the second instance, will be presented that one should rather speak about the ecology of man than that of language. Insofar as languages are not autonomous organisms but constituents of "human ecology", it seems more appropriate to highlight the ecological properties of verbal means from the viewpoint of disciplines that study factors influencing the life of language "knowers" and "doers" or language speakers and interpreters.

While focusing on people, one should treat their communicative behavior as the linguistic properties of communicating individuals that aggregate into particular societal ecosystems at various levels of social groupings, phylogenetic, professional, ethnic, cultural, confessional or economic, etc. Thus, on account of the existence of various forms of interactions, it is suggested, in consequence, to investigate the societal ecosystem within the scope of the so-called ecology of discursive communities in relation to their constitutive elements as parts of linkage systems, individuals playing certain roles of participants in group communication, verbal means, channels and communicational settings.

Postulated in the human-centered framework, the "ecological grammar", as a linguistic network formed within an ecosystem of communicating people, is counterpoised to that of "universal grammar". The latter is based on the claim of rationalist philosophy that human languages reflect extra-linguistic reality in a similar way, so that it is possible to deduce elements and structures that are primordial to human thinking from all the hitherto described languages of the world. On the contrary, the idea of ecological grammar is based on the conviction of experientially minded practitioners of communication sciences who have noticed that the manifestation forms of verbal meaning bearers are unequally put into use by communicating individuals and social groups and appear to be polymorphous when formed in dependence on their environments. It is based on the assumption that verbal and nonverbal forms of communication occur on various organizational levels of society in a twofold manner, namely, as relatively changeable practices, and also stabilizing patterns of interpreted discourses.

1. On the notion of ecology and ecologism as an investigative attitude in the sciences of language

The idea of the ecology of man, as a speaking animal whose properties are inherited biologically and transmitted culturally through generations, is put forward against the background of the ecology of language. It opposes the view of traditional linguists (cf. Wirrer 1997: 155ff; discussed in E. Wąsik. 1999b: 10ff, 41–42; and 2003: 254) who locate the lexicon and grammar of a given language within a more complex system at a higher level, the so-called language *ecosystem*, including language speakers and hearers along with their behaviors and attitudes in environmental settings. Bearing in mind that only people who communicate are accessible in the real world, the author proposes, after Franciszek Grucza (1983) and Victor H. Yngve (1996), to depart not from language, as an abstract system, but from concrete manifestations of people's linguistic properties. Thus, in such a human-centered view (postulated in E. Wąsik, 2000 and 2003), the linguistic properties of communicating individuals, who assemble into discursive communities through the realization of shared tasks, are to be treated as constitutive aspects of human ecosystems formed in the natural and socio-cultural domains of their life.

An ecological way of reasoning was rooted in the naturalist heritage of Ernst (Heinrich Philipp August) Haeckel ([1866] 1988), a German biologist and philosopher of evolution. Cf. his statement:

> Unter Oecologie verstehen wir die gesamte Wissenschaft von den Beziehungen des Organismus zur umgebenden Aussenwelt, wohin wir im weiteren Sinne alle Existenz-Bedingungen rechnen können. Diese sind theils organischer, theils anorganischer Natur; sowohl diese als jene sind ... von grösster Bedeutung für die Form der Organismen, weil sie dieselbe zwingen, sich ihnen anzupassen. Zu den anorganischen Existenz-Bedingungen, welchen sich jeder Organismus anpassen muss, gehören zunächst die physikalischen und chemischen Eigenschaften seines Wohnortes, das Klima (Licht, Wärme, Feuchtigkeits- und Electricitäts-Verhältnisse der Atmosphäre), die anorganischen Nahrungsmittel, Beschaffenheit des Wassers und des Bodens etc.
> Als organische Existenz-Bedingungen betrachten wir die sämmtlichen Verhältnisse des Organismus zu allen übrigen Organismen, mit denen er in Berührung kommt, und von denen die meisten entweder zu seinem Nutzen oder zu seinem Schaden beitragen. Jeder Organismus hat unter den übrigen Freunde und Feinde, solche, welche seine Existenz begünstigen und solche, welche sie beeinträchtigen. Die Organismen, welche als organische Nahrungsmittel für Andere dienen, oder welche als Parasiten auf ihnen leben, gehören ebenfalls in diese Kategorie der organischen Existenz-Bedingungen." (Haeckel 1866/1988: 286, the description and the opening part of the quotation in English is provided also by Nöth 1998: 332 and 2001: 78).

The term *ecology*, initially referred to the studies of the relationships existing between organisms and their environments, and has lately become popular in the domain of sociological studies concerned with the spacing and interdependence of people and institutions. Initiated by Robert Ezra Park, and Ernest Watson Burgess (1921) the idea of "human ecology" was propagated later in the works of Amos H. Hawley (1950). Additionally, the word *ecology* connoted also the care for endangered species or for the purity of the environment, in which all living systems function or live. In the 1970s, practitioners of linguistic sciences put into use the term *ecology of language* (Haugen 1972; cf. Wąsik, Z. 1993a, 1993b and 1997; Wąsik, E. 1999a and 1999b) with reference to a neutral understanding of the German term *Ökologie* coined in 1866.

Slightly earlier, however, in Leo Zawadowski's (1966) linguistic theory, the so-called ecological-relational properties of language were distinguished as opposed to lexical-relational, lexical-inherent and grammatical-inherent properties of language as a system. Zawadowski considered as unimportant for the essence of a natural language that it has an international range or is limited to a local vernacular.

For Einar Haugen (1972), conversely, the basic part of the ecology of language constituted human minds of monolingual, bilingual or multilingual individuals who take part in the interactions between members of speech communities. From that time on, external conditionings of verbal forms of communication, the dependence of languages on the people who speak them while realizing their communicational tasks appeared to be the most important factors for practitioners of language sciences. Contacts between languages and contextual properties of the manifestation forms of language become realized (materialized) exclusively through the language bearers in interpersonal communication, and, what is more, thanks to their activity.

Under the influence of Zawadowski's and Haugen's conceptions, "the ecology of language" was understood as a domain dealing with extra-systemic conditionings of language (cf. Wąsik, E. 1999b: 10ff, 41–42, see also Wąsik, Z. 1993a, 1993b and 1997). Linguistic contributions to the model of the ecological description of languages consisted in a survey of those external factors that surround natural languages and influence them, i.e., all social and cultural forces, which shape the life of individuals or communities of language "knowers", language doers and language speakers and interpreters. For example, ecological variables were taken into consideration, which resulted from knowledge elaborated in the domains of the neighboring disciplines of linguistics and the non-linguistic sciences of language. Among them are, e.g., name, history, users, territory, standardization and codification, domains of language use, symbiosis with other languages and with other semiotic systems created for the purposes of interpersonal understanding in contact situations, forms of struggle for independence, language loyalty and ethnic solidarity, legal status and attitudes toward language.

Elaborating a descriptive model for external linguistics, I have noticed (Wąsik, E. 1999a, cf. mainly 1999b: 12–14, 1999c: 57–59) that some ecological variables are more or less distant from language as a system of verbal expressions. This conclusion has entitled me to detach linguistic descriptions of languages and other communication systems from the characteristics of speaking individuals and communities, and their surroundings. Accordingly, I have postulated a distinction between (1) the metalinguistic ecology, including what can be said in linguistic sciences or in a language about (2) the ecology of language bearers, and (3) the ecology of language communication (as shown in Table 1).

TABLE I. ECOLOGICAL VARIABLES IN THE DESCRIPTION OF LANGUAGES

I. The metalinguistic ecology of a language

(1) The position of a language in linguistic classifications: language family, Sprachbund; language group; pidgin, Creole, natural or artificial language; living, developing, endangered, dead language, etc.

(2) The name of a language and language bearers: native or foreign, acquired or imposed; neutral or marked; motivated or unexplained, etc

II. The ecology of language bearers and language users

(3) Present demographic characteristics of speech communities:

– The statistical quota of native and foreign language speakers: major and small languages, linguistic majorities and minorities;

– Sociostratigraphic data of language bearers: age, sex;

– Socio-ethno-economical distribution of language bearers: individuals, groups, social classes, professional and/or confessional communities; nation or nationality; sedentary or nomadic, consolidated or Diaspora, etc.

– The question of anthropological properties of language bearers;

(4) Territorial, geographical and political settings: compact or diffuse; indigenous or alien; language island, state, country, region, district, etc.

(5) An external history of language and language bearers: conquests, expansions, peoples wanderings or migrations; formations and impacts of power centers, empires, national-liberation movements on the ground of language, etc.

(6) Attitudes towards language:

– Language as a criterion of ethnic identity and other semiotic systems characterizing language bearers: language and literature, anthem(s), flag(s), costume(s), music, folk dance, artifacts, etc.

– Language loyalty and ethnic solidarity: acceptance or rejection of the language status; fidelity or renouncement; intimacy, proximity or alienation on the ground of common language among participants of interpersonal communication, inter-group or inter-ethnic; the sense of domination or subordination; the evaluation of superiority, inferiority or equality of a language used as a native or foreign means of communication (reciprocally or unilaterally), etc.

(7) Language policy and language planning:

– Standardization, codification, autonomization and maintenance of the vitality of a language: cultivation of a language, implementation of a language for official matters; language norm, orthography and orthophony, unification of writing and pronouncing or acceptance of variants in spelling and, speaking; grammars and dictionaries, etc.

– Organizational and political support for a language and forms of struggle for its maintenance, cultivation and education: formal or informal, parliamentary or revolutionary; manifest or hidden, tolerated or forbidden, etc

III. The ecology of language communication

(8) The media-related realization of a language: types and styles of textual messages and forms of communication channels; prose and poetry, folklore and proverbial phraseology; metaphorical idioms and world view; vocal-auditory, written, printed, visual or palpable; dialogical or monological, direct or indirect, official or casual, spontaneous or formal; frank or reserved, informative or performative; scientific, artistic, colloquial, or professional; urban, rural, metropolitan, or regional; journals, newspapers, radio, television, etc.

– Language varieties: dialects, local vernaculars, contact varieties, functional styles, professional jargons, registers, etc.

(9) Domains and functions of language use: intrapersonal vs. interpersonal: dyadic, small-group, public and mass communication; family, market, country fair, shop, school, church, theater, carnival, stadium, office, court, army, etc.; stable, temporary, compulsory, facultative, additional, complementary, progressive, declining; integrating, separating, symbolic, referential; acquisition, use, attrition, etc.

(10) Symbiosis or conflict with other languages in contact situations: bilingualism, trilingualism, multilingualism or diglossia, triglossia, multiglossia; substrate, superstrate or adstrate; borrowings, cultural transfers and language interference; boundaries of language, religion, states, powers, etc.

In the investigative practice of linguists, the term *ecology of language* was referred mainly to the extralinguistic properties of languages understood as systems of verbal expressions spoken and understood in human communication, and sometimes it could have a broader range (cf. Wirrer 1997: 155ff; and Di Cristo 2000: 19f). In the latter case, the system of language was seen as embedded within a more complex system at a higher level of the so-called language ecosystem, including grammar and lexicon and its environment, the people who communicate with their political settings and their behavioral properties and attitudes. Attention then was paid to the processes of changes within a language ecosystem as a whole in order to answer how the environmental factors influence the functioning of elements and structures of verbal means of communication.

The notion of ecology with reference to language as an object of linguistics proper as well as to other objects studied by semiotic-communicational disciplines was indeed metaphorical. Its use resulted from the treatment of language or other systems of signification and communication as an autonomous agent or living subject. However, one should bear in mind that the only active subjects in linguistic communication are people. Hence, for the aims of the theory of human communication it would be more appropriate to speak rather about the ecology of man and society than that of language and culture.

2. Specifying the concept of an ecological grammar of linguistic linkages realized in human communication

Since investigative objects of linguistics proper, are not organisms but rather constituents of "human ecology", I have tried to reconsider the ecological properties of verbal communication in the light of disciplines that study the spacing and interdependence of people and institutions. So to speak, I have departed from the assumption that the basic parts of the so-called "ecology of language" encompass human minds of monolingual, bilingual or multilingual individuals who are engaged in interactions with other individuals. Departing from the belief that communicating individuals constitute the main object of linguistic studies (in accordance with Victor H. Yngve, 1996), I have confined myself to treating the communicative behavior of people as observable links that mediate between linguistic communities and their surroundings within a span of years, in a certain territory, in a given country, and/or in the relationship between states. In such a human-centered framework, communicational forms of interpersonal linkage systems are considered as discourse genres or patterns recognized by communication participants as belonging to certain types of discursive practices. Moreover, it is assumed that the linguistic properties of communicating individuals are to be searched in discursive ecosystems aggregating at various levels of human communication. Considering the postulates of semioticians (cf. Enninger, Wandt

1984: 29ff, mainly 32) the societal ecosystem might be also within the frames of the so-called ecology of sign, including not only interlocutors but also all communicational phenomena.

Nonetheless, within the framework of human-centered linguistics, these constitutive elements of linguistic ecosystems might be considered as parts of linkage systems, individuals playing certain roles of participants in group communication, props, channels and communicational settings. As regards the verbal forms that unite individuals into communicating groups through different channels of communication, they are studied along with nonverbal behavior in the realization of linguistic properties of people, hitherto included in the realm of bio- and anthroposemiotics.[1]

Discussed in the context of human sciences, parallel to the newly launched "communicational grammar" (postulated, *inter alia,* by Geoffrey Leach 1983), the conception of "ecological grammar" refutes the idea of "universal grammar" advocated by the followers of the distinction between transformational-generative and traditional grammar. The search of the latter for universal grammar is rooted in the hypotheses of rationalist philosophers that all human languages reflect extra-linguistic reality in a similar way. Henceforth, some philosophically inclined linguists, *inter alia,* Anna Wierzbicka (cf. 1972), have believed in the possibility of deducing from all hitherto described languages of the world the elements and structures that form primitive constituents of human thinking. On the contrary, ecological grammar has grown out of the experience of the practitioners of human sciences bearing in mind that the manifestation forms of verbal behavior of communicating individuals and commmunities are polymorphous and unequally put into use when formed in dependence on their environments. It is based on the assumption that verbal and nonverbal forms of communication occur on various organizational levels of society in a twofold manner, namely, as changeable practices and stabilizing patterns of interpreted discourses.

---

1 To be mentioned is a conviction of some philosophers (e.g., Searle 1983, 1992) and biologists (Kull 2000) that between human beings and other organisms in the living world there is a biological continuity in evolution. And, such properties as the possession of consciousness, intelligence and the faculty of language, the aptitude of rational thinking, etc., are seen as phenotypic features of an organism resulting from the interaction of the genotype and the environment (cf. Dawkins 1982 and Wąsik, Z. 2001a: 85).

In this context, it would be appropriate to mention that the partial inspiration source regarding the sole usage of the term *ecological grammar* was the abstract of the paper published by Albert Di Cristo (2000), who departed, while referring to the book of Knud Lambrecht (1994), from a language-centered perspective. For the purpose of comparison, it might be relevant here to quote Di Cristo (2000: 19–20) in original:

> Dans cette perspective, nous proposons d'envisager comme cadre interprétatif de la prosodie celui d'une *Grammarire Ecologique*, l'expression elle-même étant empruntée à Lambrecht (1994). Avant d'aller plus loin, il importe de préciser que le term de grammaire n'est pris ici dans son acceptation restrictif de grammaire formelle (bien de la grammaire formelle puisse et doive, selon nous, être une composante majeure de la grammaire écologique), mais dans la signification extensive de : description des mode d'existence et de fonctionnement d'une langue naturelle où, éventuellement et plus largmement, de toute sémiotique. Le term *écologique* que nous associons à celui de grammaire tire sa légitimité de l'axiome selon lequel le language est un éco-systeme, c'est-à-dire un mode d'expression qui s'adept en permanence au milieu dans lequel il se déploie, principalement en fonction de l'environment cognitif versatile de ses utilisateurs et des pressions exercées par les fluctuations constantes de la force interactionnelle qui régule les échanges conventionnels. Telle que nous la conservon, La Grammaire Ècologique s'incrit donc a la fois dans le paradigme d'une théorie pragmatique large de la communication (incluant des aspects qui relevent de l'illocutoire de l'énonciatiation, de l'interaction, de la contextualisation et de l'expression de l'affect), et dans celui des sciences cognitives ...

Entering the domain of "ecological grammar" from a human-centered perspective, I have been more interested in the methodological consequences of this new interpretative framework of communicational studies related to verbal discourse and sociological pragmatics rather than to launch the new idea of a grammar of a certain language. Having a certain experience in the domain connected with the ecology of language extending beyond the boundaries of linguistic studies, I am aware that the term *ecological grammar* affords the possibility of embracing in one approach both systemic as well as ecological properties of language. On account of the fact that the paradigm of ecological thought is so well developed at present,[2] practitioners of human-centered disciplines may feel truly enti-

---

[2] Worth mentioning here are especially the works of Bateson 1972; Barker 1968, Barker, et al. 1978; Makkai 1993, Kull 1998a, 1998b, 1999, 2000, Nöth 1996, 1998, Ingold 1999 [1996], Hornborg 1996. For further details, in the context of linguistic studies, see also Fill 1993, 1996, as well as Fill and Mühlhäusler 2000.

tled to settle on the primacy of biological, cultural and psychological approaches to communicating individuals as organisms, as persons and participants in social roles.

Having worked towards redefining the concept of the ecology of verbal discourse from a human-centered perspective, I came to the conclusion that this domain should encompass the interrelationships between linguistic properties of individuals and groups with their surroundings, in which they function as communicating individuals, and to which they refer their expressions.

Since individuals constitute a group of communicating people they are characterized, *inter alia,* by such changing properties, as, e.g., (1) statistical quota and distribution, (2) division into sex and age groups, (3) normative acts of communicational conduct and their execution in a language (4) cultural patterns and forms of organization, (5) attitudes towards insiders and outsiders, and (6) internal and external contacts and behavioral codes among the communication participants (cf. Haarmann 1989: 173f). In dependence on the domain of control, the context of situation, the social stratification and communicational tasks of the participants of social communication, they may be considered as having an impact upon the differentiation of verbal behavior.

3. Presenting social groups in terms of ecologically determined dynamic systems

Having elaborated the idea of ecological grammar, from a human-centered perspective, which reconstitutes certain language-centered models (e.g., that of Di Cristo, 2000, quoted above after E. Wąsik, 2003: 258–259), we do not then focus on relationships between verbal sound-chains, but rather on communicative linkages between participants in social actions, which are in turn influenced by ever-changing, ecological conditionings. Thus, the "ecology of verbal discourse" is seen as including the properties of communicating individuals and groups of people and the properties of their surroundings, in which they function as organisms and as participants in social roles, and to which they refer their means of expressions in interactive and interpretative activities.

The use of a human-centered framework enables researchers to treat communicative linkages in a dynamic way, because individuals as parts of constantly changing social groupings are dependent upon the ecological variables which determine the modes of their existence and formation into relatively self-governing entities. As far as lower-order linkages are often subsumed within those of a higher order, any ecologically determined linkage may thus be described as developing and becoming more or less autonomous from any point of view independently of whether it is incomplete or complete, while focusing on the nature of its physical constituents. Interpreted and described in terms of sociology and communication sciences, the relations between communicating individuals can serve as a basis for distinguishing various types of interacting groups, classified with respect to their communicative properties. For example, a typology of communicative linkages, or "ecolinkages", might consider both hierarchy and inclusiveness in terms of distance; stronger linkages allow for those which are further from the core of society, and which thus crystallize as "autonomous agents" (see E. Wąsik 2003: 261). One could examine the direction in which long-lasting communicative linkages develop; first becoming homogeneous communities and then societies with permanent bonds. In so doing, one could observe what types of interacting subordinate groups (along with their particular spheres of influence) evolve in time and space, eventually becoming ethnic, national, religious, professional or natural and cultural ecosystems.

Based on empirical techniques or investigative methods elaborated by sociologists, we usually obtain a static community-related image of a given group of people. However, the application of the perspective of human-centerd linguistics demands from us to present multilingual linkages in a dynamic way. Following sociological usage, one may distinguish between communicative linkages and communicative communities. The first type of social grouping, linkages, discussed by Jeremy Boissevain (1974), is based on temporary or long-lasting interactions in a diffused way; in totality, they might be considered as a set of collections. Communities, however, constitute a set of members who are mutually concatenated by common organizational principles. Communities, being human institutions, should be seen as 'collectivities', not just 'collections' of individuals, in accordance with with Graham MacDonald and Philip Pettit's view (1981: 107f; quoted in Downes 1998 [1984]: 106). Due to a collectivity principle individuals and groups, which will be of significance for human-centered linguistics, may be characterized by at least fifteen features (adapted with slight modifications from: Downes

1998 [1984]: 107): (1) changeability of membership, (2) continuity in time, (3) disseminated participation in different groups, (4) physical discontinuity of memberships, (5) direct or indirect assemblage of group members, (6) temporary (evanescent) or long-lasting (enduring) membership, (7) hierarchical or egalitarian stratification, (8) solidarity and self-consciousness of belonging to particular groups, (9) identity sense of exclusion or inclusion in group membership, (10) reference groups, (11) variability in group practices, (12) exertion of normative pressures and sanctions, (13) centrality and/or peripherality in group memberships, (14) differentiation in internal structures, (15) stigmatic or charismatic properties in stereotypical categorizations. Those features may be explained in the following statements unconnectedly quoted after Downes (1998 [1984]: 107).

(1) "A group remains the same group although the individuals who constitute it may change." (2) "A group remains the same group as it continues in time, and in spite of changes in membership." (3) "The individuals who constitute a group may at the same time be members of other groups." (4) "The individuals who constitute a group need not be in physical proximity." (5) "In primary groups, the members are directly related in face-to-face interaction. In secondary groups, the members are indirectly related, according to some wider criterion." (6) "A group may endure for generations, or may be evanescent, forming for a particular purpose or circumstance and dissolving afterwards." (7) "Groups may be stratified relative to other groups in a hierarchy from higher to lower according to some evaluative criteria." (8) "Individuals can be grouped together on any criterion. Not all criteria are equally salient. The most important criterion is the self-conscious significance of the group – how it views itself as significantly different from other groups, and how solidary its members feel." (9) "An individual's multiple group membership make up their social identity." (10) "An individual may find identity or values in those of some group of which they are not a member." (11) "Practices are regularities of behavior, which characterize groups." (12) "Pressures are exerted on a groups's members to make them conform to the practices of the group." (13) "Individuals may be central or peripheral members of the group." (14) "Groups have different kinds of internal structure. Some are integrated, some diffuse. Some are informally organized while others are institutions, with formally specified roles and statuses." (15) "Groups may be characterized in terms of stereotypical properties

and members stigmatized or granted prestige in terms of objectively false categorization, e.g., *Blacks, immigrants, women, youths, Irish*, etc. (see chapter below)."

Accordingly, our knowledge of communicational processes within ecosemiotically defined groups has had to be supplemented by considering the possibility of changes. Groups of people studied by human-centered linguistics are not static in reality, but dynamic, and their existence is conditioned only by communication among their members. For example, the number of speakers within a traditionally distinguished speech community is assessed in a static manner. When taking into account the number of interacting linkages, statistical data always had to be modified.

The linguistic properties of people constantly change because individuals as persons participating in social linkages as assemblage groups are dependent upon biological, psychical, social, cultural, technical, political, economic, and other ecological conditionings, which take part in the determination of the modes of their functioning and their development into communicative-interactive entities.[3] Because the linkages of a lower order are usually situated within the linkages of a higher order, the sociological notion of autonomy appears to be appropriate with regard to the self-government of a small-group community applying its own laws and functioning within the larger structures of a given society.

As a consequence, any ecologically determined linkage might be observed as developing and becoming more or less autonomous from any point of view independently of whether it is focused or complete in character. Interpreted and described in terms of communication sciences, the relations between communicating individuals can serve as a basis for the distinction of various types or kinds of interacting groups in terms of communicative properties. Thus, a typology of such communicative "ecolinkages" has to consider the degrees of their discreteness, peculiarity, separateness, independence, self-existence, and self-reliance. In this set of the six qualifiers of autonomy, following Stanisław Pietraszko (1992, discussed in Wąsik, Z. 2000: 32), one can notice both a hierarchy and inclusiveness: less distant, stronger linkages include those which are

---

[3] In this context, to be exposed is the opinion of Geertz (1996 [1963]) who distinguished among primordial ties linking social communities, as, e.g., blood kinship, race, religion, customs, also language as a factor forming the center of social matters.

weaker and more distant from the core of society crystallizing as an autonomous agent (cf. Wąsik, Z. 2000: 32). One may examine the direction in which the development of long-lasting linkages goes so that they become homogeneous communities with permanent bonds. This means, one can find out what types of subordinate groups interacting with each other evolve in time and space along with their particular domains of control into ethnic, national, confessional, professional or natural and cultural ecosystems. The latter groups might be considered, following Ludwik Zabrocki's distinctions, as communicative communities, which may be divided into: "(i) active and passive, (ii) durable and non-durable, (iii) loose and compact, (iv) primary and secondary, (v) superordinate and subordinate" (see Bańczerowski 2001: 38). In Zabrocki's view, each individual is, simultaneously, a member of various communities determined by communicative domains of life, as, e.g., family, home, work place, church, political organization, etc. Each community, in turn, is determined by extra-communicative factors of heterogeneous nature depending on the geographical, economic, political, ideological, and cultural conditionings of environment, in which communicating individuals as members of societal groups live.

4. Linguistic ecosystems under the pressure of globalization and bioinvasion

Having outlined broadly the concept of ecological grammar as a human linguistic network formed within an ecosystem of communicating people, we could attempt at formulating some statements about the impact that a global language of world-wide-web communication may have upon local languages. As such, these statements might be referred to global English and other languages spoken in the world by local non-English communities as well as English transmitted on a global scale and the English varieties understood on a local scale. Specifically, we would be interested to learn how the significative function of a global language may be influenced through the interpreting activities of members of local linguistic communities. We had also to take into account the interference of indigenous linguistic distinctions on the significative and interpreting activities of local individuals participating in a global network of communication. In selecting the subjects of our human-oriented study, we would look over the linguistic behavior of such participants of global communication who represent certain local interest groups or professional enterprises though the exchanges of letters, transmitted by surface or electronic mail in English. Another type of informants would embrace those individuals who take part in the production or reception of advertisements in journals or billboards of foreign companies that strive for the domestication of their messages in verbal means of local communities. The list of investigated subjects participating in various global and local linkages could be extended through the enumeration of communicational domains and their ecological settings in detailed field explorations.

Accordingly, we would concentrate on exploring the "foreignization" processes of discursive practices of people who participate in longer lasting English linkages. Working in the field of human sciences, we might be interested to explore a wide range of ecological factors that affect the dynamics and change of the linguistic properties of communicating individuals intrinsic to their nature as organisms and their role as communication participants at a global scale. These studies might exhibit such external factors that sustain or threaten the vitality and maintenance of hitherto-existing linguistic communities in their local environments. Being interested in any aspect of communicating individuals, we would focus on cross-disciplinary questions, regarding the biological and cultural diversity of linguistic communities against the background of ecolinguistics, contact linguistics, personal identity studies, language policies and the linguistic rights of people.

As can be observed, this preamble has reduced the scope of the term *globalization*, discussed by sociologists (see Sztompka 2002: 510, 581–597), among others, Roland Robertson (1992) and Anthony Giddens (1990), as well as the theoreticians of mass media sciences, mainly Marshall McLuhan (1964), cited by Marcel Danesi (2002: 12, 142f), to the linguistic communication of people. Worth mentioning are some constituents of this "set of processes that create the social world as a unity": (1) mass media and Internet, (2) supranational associations, organizations and social movements, (3) economic and financial interdependencies between local communities, and (4) human mobility in the division of labor around the world. As most relevant for us is the world-wide-web of computer connection between various parts of the globe and the verbal means of English transmitted and received in speech sound and/or writing by communicating individuals, as, e.g., pilots, controllers of flight area, tourists, scientists, computer programmers, businessmen and diplomats.

Be that as it may, the expression forms of verbal means are not to be equated with natural or cultural goods being transferred or exchanged, or shared as commodities in a worldwide trade. In fact, commodities leave their source location and arrive at a target location, or eventually loose their source value and acquire a new target value, while verbal means, by contrast, which are transmitted from a source to a target location, constitute the physical part of the domain of human expression. While remaining in the logical domain of a source agent, the material shape of verbal means, as particle and wave duality, is received and (or not), by a target agent. As such the logical domain of communicating people exists separately in the knowledge of both the source and the target agents as a mental relationship between the two associated domains, the domain of expression and the domain of reference. Thus, individuals communicating about the same domain of reference are supposed to be endowed with the knowledge of how to interpret the domain of expression of a given language in a relatively similar way. This referential knowledge might be possible, when similar products, patterns of behavior, values, and ways how to satisfy individual needs and social requirements are distributed on a global scale. Moreover, one has to assume that not only the content of interpersonal communication but also the basics of scientific and school education in every part of the globe are comparable. From the viewpoint of the ecological conditionings of human activities in the interpersonal understanding processes, this fact is not always true. Thus, taking into account the role of individual experience, our attention should be directed towards the biological concept of "embodied semantics" derived from the interactions between organisms and their natural and cultural environments. That is, we would encompass the subjective knowledge of humans gained through sensory actions in cognition "embodied in the lived histories of organisms, their communicative, cultural and linguistic practices" (Foley 1997: 177). Accordingly, we would especially consider the approach to meaning and understanding, related to the subjective enactment of the world in sensory cognition, which has been proposed by Humberto R. Maturana and Francisco J. Varela (1987).

Uniting two orders of human communication, culture and nature, in one investigative attitude, we might share a conviction of contemporary philosophers, e.g., John Roger Searle (1992) that between human beings and the remaining constituents of the living world obtains a certain continuity (cf., *inter alia,* Kull 1998, 2000, Ingold 1999 [1996]). Hence, such species-specific properties of humans as consciousness, intelligence and the faculty of language, the capability of categorization, the aptitude of rational thinking, etc., would be considered as the biological phenotypic features of an organism resulting from the interaction of the genotype and the environment (cf. Dawkins 1982). They might be seen as products of biological evolution, in the same way as all other phenotypic features. It is probable that biological processes have caused the emergence of self-consciousness, which forms a part of the natural biological order just as the other biological phenomena are, such as photosynthesis, metabolism or mitosis, and the like (cf. Edelman 1992).

Sharing the belief of biologically informed semioticians, we might be tempted to reject the assumption that languages as a domain of significative means are only culturally transmitted through generations (for details see Wąsik, Z. 2001a). They are probably shaped also by the linguistic properties of people as organisms in particular along with species-characteristic faculties of man in general, being genetically inherited. Cultural transmission in education and communication, both on a local and global scale, is connected with human conventionality and creativity contributing to the formation of new inventories of verbal means, and their functional and stylistic varieties, allowing us to realize communicational tasks of people never registered before, etc.

Since humans as organisms are constituents of natural ecosystems, which change as a result of their mobility and interactions with and within changeable environments, globalization on the cultural level is also accompanied by bioinvasions on the natural level, as it is argued by Claus Emmeche (2001). Globalization as a process is usually specified as the transgressing expansion of the modern way of life, which occurs at the expense of the more traditional forms of life. By immigration, relatively homogeneous societies in their demographic and cultural substance transform into multicultural societies. In the context of nature conservation, bioinvasion due to globalization may cause the breakdown of natural barriers between human ecosystems. When *homo transportans* is travelling around the globe, he may influence the core of the evolutionary processes in the natural ecosystems on the local scale. As such, this may pose a threat to communities of animals, plants, and microorganisms including their interactions with their organic and inorganic environment that contain a diversity of species not found earlier in a variety of some other ecosystems, or restricted to a few or one specific kind. Bioinvasion is thus understood as "the (intended or non-deliberate) introduction by human activity of non-native species into ecosystems in which such species have not been found before, and in which they tend to become invasive, that is, they spread, colonize and become established, often at the cost of the distribution range of some of existing species" (after Emmeche 2001: 240).

As we may conclude from Emmeche's reasoning (2001: 239–258): (1) Bioinvasion on a global scale can contribute to the perturbance in the balance between the rate at which new species are formed and the rate at which some old species become extinguished. (2) Increased bioinvasion will globally decrease biodiversity and disturb the evolutionary dynamics, which takes place in local ecosystems. (3) Bioinvasion, attributable to human activity, can cause an intentional or unintentional damage to an ecosemiotic system having its own life and maintaining its integrity in the coherent complex of organic, cultural and social relations.

After all, bioinvasion due to globalization in human ecosystems should not only be perceived as a threat towards the local stability of local flora and fauna. It might be also consequential in the creation of new ecological grammars in which indigenous and foreign domains of cognitive reference intermingle, i.e., in the formation of new linguistic communities based on temporary and long-lasting linkages of communicating individuals.

Chapter Five

Sociological aspects of linguistic pragmatics in the light of hard-sciences

The subject matter of our discussion constitutes the investigative consequences in the study of verbal communication which might particularly result from a rigorous application of the hard science perspective postulating at the same time to discard the validity of the statements that derive their substantiation from the fields investigated by soft sciences. Against the background of the distinctions between the physical and logical domains proposed in the so-called human linguistics, which include both observable and concluded facts as extraorganismic and intraorganismic properties of communicating individuals, it is emphasized that the philosophical foundations of pragmalinguistics are indispensable for human-centered pragmatics considering the self-awareness of communicators not only as intellectual and emotional activities of the mind but also as chemical-electrical and motional-kinetic activities of the body.

1. Physical and logical domains in the investigative field of human linguistics

Having rejected the distinctions provided by soft-sciences, both scientists and ordinary human beings would be unable, due to the lack of theoretical constructs forming the basis of solipsistic experiences of observers, to communicate about the nature of things and states of affair that are remote in time and space. The only thing they could state about the real world of communicating people is that there are observable links between individuals forming parts of a dynamic linguistic community with open boundaries. These inter-individual links constitute energy flows exchanged through verbal expressions. However, as it was said before (cf. p. 88), linguistic expressions cannot be identified with natural or cultural goods being transferred exchanged or shared in the manner of commodities. As far as commodities are transmitted in totality from their source to their target location, verbal means being sent from a source location arrive only with its physical part at a target location. Their mental counterparts remain in the memory of a source agent. Being received, conversely, they evoke similar images in the mind of a target agent. Due to the substitutive character of linguistic means the logical domain exists separately in the knowledge of both the source and the target agents as a mental relationship between the two associated domains, the domain of expression and the domain of reference. As a result, two individuals who communicate must be endowed with approximately the same knowledge about the domain of their reference in order to understand each other while interpreting the respective domain of their expression in a more or less similar way.

2. How to do things with "sound waves" or with "words"?

To begin with, the point of our departure will be human linguistics which has been specified by Victor H. Yngve (1996) as "the linguistics of people" in opposition to "the linguistics of language". In the investigative field of human linguistics, the subject of a scientist's interest encompasses those linguistic properties of communicating individuals that are relevant for the realization of their communicational tasks considered on the one level as real persons, and on the other as participants of social communication. From such a viewpoint, the linguistic phenomena are localized as observable properties of people within the physical domain. Therefore, human linguistics is assumed to be a scientific discipline which focuses on concrete people, and not on an abstract language. Following Yngve's (1996: 4) opinion: "We find in nature only the physical waves; their interpretation is entirely in the heads of the speakers and hearers. A scientific analysis must include, besides the study of the physical sound themselves, a careful and detailed study of the people who produce and interpret the sounds and what they are doing at the time. As scientists, we would also like to understand the source of the compelling illusion that utterances and the parts of utterances do exist in nature." Only the physical domain (including, apart from people and sound waves, also the context comprising some other corporeal objects) can be testified by "hard science" and therefore constitutes an exclusive source of knowledge about interpersonal communication. The concept of the physical domain is counterpoised to that of the logical domain. In Yngve's view, the logical domain, examined by "soft science", belongs to the investigative realm of philosophy, logic and linguistics. As he claims, traditionally inclined linguists dealing with language in semiotic and grammatical terms place their object of investigation in the logical domain, which is not testable by experimental methods. Accepting the hard-science tenet that the only accessible objects of scientific study, understood in terms of physics, chemistry and biology, are the linguistic properties of human individuals communicating with other individuals in temporary and long-lasting linkages, we might agree with the opinion that "[t]here is no such thing in nature as an utterance that carries with it a linguistic segmentation or structure of any sort, whether in terms of phonemes, syllables, words, sentences, or any other of the constructs usually invoked to describe them" (1996: 9). However, we cannot say the same about the next statement: "Instead we have in nature only the physical sound waves themselves and the people producing, sensing, and interpreting them" (1996: 9). In fact, only the first phrase pertaining to observable channels and referring behavior of communicators is true. Without a doubt, "interpreting" activity has to be included into the logical domain because it is based on inferences and conditional reasoning, in the same way as "competence", which is seen by the author himself as "not a part of the real world" (Yngve 1996: 341, cf. also 97). Moreover, Yngve (1996: 189ff) has introduced into his theory the term *task* to "understand" the aim of people's communicative behavior. He speaks even about the *task hierarchies* and *subtasks* that are executed by the participants of communication who act as members of different groups. The notion of "task", connected rather with the investigative area of psychology, along with notions of "knowledge", "concepts", and the like, deviates also, as a matter of fact, from the terminological and methodological assumptions of human linguistics as a hard science.

Those who accept Yngve's distinction between the two domains – the physical domain and the logical domain – to the interpretative framework of linguistic pragmatics are undoubtedly aware that communication is an observable concrete fact only when people talk to each other producing and receiving verbal means. However, the occurrence of mutual understanding between communication participants is a concluded mental fact. Considering the ontological status of communication constituents, what appears to be observable in an actual speech act are, firstly, the people who communicate, secondly, the physical sound waves linking the speaker(s) and the hearer(s), and thirdly, all those physical objects and parts of the surroundings which communication participants utilize as relevant for the realization of their tasks. To summarize, one can say that at least three elements constitute an observable whole being empirically available and proved, namely, the communication participants, verbal products as material bearers of meanings and the situational context. What the people mean, what is the referential value of their communicational means, cannot be known without inferences based on logical reasoning.

According to the assumptions of human linguistics, one observes the linguistic properties of particular persons, relevant for these persons communicative behavior. Such properties are, e.g., being bilingual, having a certain regional accent, knowing how to refer to different people in different contexts, knowing which referring contexts are current, knowing how to ask a question, producing the sound of the word *why*, being on the spot to answer a question (or not) or being engaged in the task of answering a question (cf. Yngve 1996: 123–124), but only some of them are adequate in a particular situation. In the interaction, there are always more persons than one who are involved in the realization of communicational tasks, so that people are doing things through conversation (like buying and selling, haggling over a price, arguing, convincing, discussing, reaching an agreement, coordinating work on a common undertaking, asking and explaining, teaching and learning, disagreeing) usually cooperatively (cf. Yngve 1996: 84). They interact in groups (such as family, work, sports groups, school classes, committees), in groups coordinated by telephone calls, exchange of written communication, in groups involving publication, radio, television, readers for some selected writings. As members of interacting groups, people develop certain common properties by virtue of reading, hearing, or viewing the same materials (cf. Yngve 1996: 85).

Observable from a hard science perspective group members, as real persons involved in interaction with their environments, constitute only physical objects of assemblages which exist together with the relevant sound and light energy flow (of speech sounds and the light energy associated with gestures) as well as other (non-personal) objects and places of communicative relevance. However, on the theoretical level, they are considered as communicating individuals playing the role of *participants* in (communicative) *linkages* together with other linkage constituents, called in human linguistics, respectively, as *channels, props* and *setting*. The linkages, as assumed constructs, can be adjusted by each researcher arbitrarily, being as such delimited in space and time (cf. Yngve 1996: 126ff and 231ff). There are different types of linkages, such as, small, large, brief, long-lasting, broad or narrow. According to Yngve, one can prove that interacting individuals are linked with each other and with their environment as parts of observable reality. Among observable data one can find also, e.g., composite linkages that can be directly or indirectly coupled (if they are interacting with each other without requiring any third linkage or through one or more other linkages (cf. Yngve 1996: 203).

The linkages can be coupled in some cases through their arrangement (cf. Yngve 1996: 214). One can distinguish also focused linkages, which include only a limited range of the observable phenomena, and complete linkages, which include the full involvement of the participating individuals (cf. Yngve 1996: 180).

One may study the linkages in contact situations (through covered or overlapping participants), as well as linkages interacting through channels, props and settings, which trigger communicative behavior in the linkage. In any case, one can notice that there are some linkages, which form a hierarchy of stable couples of communicating individuals, and that there are also groupings of linkages, being mutually concatenated, which always change, so that the picture of a linguistic community appears to be dynamic and not static as was traditionally depicted.

In linguistic pragmatics, developed after the proposals of the philosophers of language, *inter alia,* John Langshaw Austin (1962), John R. Searle (1969), Stephen C. Levinson (1983) and Geoffrey Neil Leech (1983), the performative function of speech acts are ascribed to utterances the function of which is to achieve certain communicational goals of individuals indirectly or directly. However, bearing in mind the assumptions of human linguistics: "It is sound waves, spoken and understood that the people ... are doing things with" (Yngve 1996: 85), and not words as was stated by John Austin in the famous title of his series of lectures, *How to Do Things with Words*, which, in consequence, should be rather understood as "how to do things with sound waves". Analyzing the example: *I bet you sixpence it will rain tomorrow*, Victor H. Yngve explains that what people are doing with sound waves (and not with words, as misleadingly claimed by John L. Austin), "involves significant changes in the properties of both the speaker and the hearer: the properties of the speaker change to reflect that he has made a bet and the properties of the hearer change to reflect that a bet has been made." As Yngve claims, "[w]e can test that these changes have actually taken place by correlating the observed presence or absence of rain on the next day with the observed passing of sixpence from one to the other. Since the bet requires two people, it can be seen as a property of the group; and it is a property of this particular group for this particular stretch of time" (1996: 85).

As a matter of fact, pragmatics as a discipline dealing with the practices of verbal communication is interested in discovering the general schemes of linguistic behavior of people, which govern the conversational rules of speaking and understanding processes in interpersonal communication. While speaking constitutes an activity of senders who transmit messages oriented towards achieving intended goals, their comprehension belongs to receivers.

Without a doubt, understanding the tasks, intentions or goals of senders appears to be possible for receivers, as communication participants or practitioners of linguistic pragmatics, only through the interpretation of observable facts on the basis of their subjective insights achieved through solipsistic introspection. In the latter case, it is sometimes intuition which plays an important role. In principle, one assumes that the domain of linguistic pragmatics comprises the search for the meaning of language utterances in relation to their authors by considering the role of their situational context, or more broadly – their social and cultural context.[1]

To be precise, while studying the pragmatics of speech, a linguist has to take into account, firstly, that the speaker may express through his utterance the facts, which stand in conformity with a certain state of the

---

1   The boundaries between pragmatics and semantics were specified by Leech (1990 [1983]: 5) as follows: (i) the pragmatic interpretation of a sentence is distinct from its semantic representation; (ii) pragmatics is principle-controlled and not rule-governed; (iii) the principles of pragmatics are non-conventional, i.e. motivated in terms of conversational goals; (iv) pragmatics relates the sense of an utterance to its pragmatic (or illocutionary) force; (v) pragmatic correspondences are defined by problems and their solutions (not by mappings as the grammatical correspondences); (vi) pragmatic explanations are primarily functional (not formal as the grammatical explanations); (vii) pragmatics is interpersonal and textual (not ideational as grammar); (viii) pragmatics is describable in terms of continuous and indeterminate values and not in terms of discrete and determinate categories.

reality communicated in a word-for-word manner, i.e., presenting a literal semantic content of what he utters in a phonetic "locutionary act".[2] And secondly, – he has to be aware that the speaker may act intentionally realizing a certain aim in view, communicational goal or task (considered also in terms of conversational implicature), providing his utterances with a certain pragmatic value, or illocutionary force.[3]

Depending on cultural and environmental factors, from conversational situations as well as from pragmatic factors, such as the types of interpersonal relationships between interlocutors, their age, gender, the degree of intimacy, the purpose of a polite exchange, the speaker chooses different communicational strategies hoping that they might be appropriate and effective (cf. pragmatic distinctions discussed in the context of lie and lying by Jolanta Antas (2000 [1999]: 250 ff). Apart from this, the effectiveness of communication may be influenced by the nonverbal and verbal behavior of interactants (such as, e.g., gestures, face mimics or

---

[2] In the theory of speech acts, the distinction between locutionary, illocutionary and perlocutionary acts has been made following to John L. Austin (1975 [1962]: *How to Do Things with Words*). The notion *illocutionary force* introduced by Austin along with *conversational implicature* developed by Herbert Paul Grice (1975) has challenged the understanding of the term *meaning* coming from structural-systemic linguistics. A further contribution to the discussion about what the meaning of an utterance is and what the speaker means through uttering a certain sequence of words has been made through the introduction of the distinction between implicit and explicit ways of communicating the meaning conducted within the relevance theory framework of Dan Sperber and Deidre Wilson (1995 [1986]) by Robyn Carston (2002).

[3] The pragmatic value of an utterance or, in other words, its pragmatic force, constitutes a product of the principles of textual rhetoric and the principles of interpersonal rhetoric (cf. especially Leech 1990: 15–17). In particular we assume that the speaker acts basing on the principle of cooperation, and expressing his communicative goals in a clear and compact manner, observing the maxims of quantity, quality, manner i and relevance, he is guided by sincerity and considers in conversations not only the principles of textual organization, but also the principles of interpersonal contact. This means, the speaker bears in mind the principle of cooperation with the four mentioned maxims and the politeness principle with the six maxims: (Tact Maxim, Generosity Maxim, Approbation Maxim, Modesty Maxim, Agreement Maxim, and Sympathy Maxim). It is to be remembered that the importance of these maxims is different in different cultures; hence, some researchers devote their attention to comparative studies of the cultural style of politeness phenomena (cf., e.g., Antas 2000: 251).

tone of the voice).[4] However, the effect of the communicational act, which is achieved in a perlocutionary way, can be evaluated on account of verbal or nonverbal reactions of addressees.

Based on the assertions of cognitive linguists, we may agree that there are no clear cut boundaries between semantic and pragmatic approaches with respect to the referential value of sentences that people use and learn in verbal communication. When, for example, a Pole hears: *Czy nie masz drzwi w domu?* [Don't you have any doors at home?], in the situation where he has entered the room without closing the door, then he habitually recognizes this statement based on his knowledge of its earlier usages not as a question 'whether someone possesses the door at home', but as an impolite demand, expressing irritation, 'to close the door'. He is not supposed to interpret the semantic meaning of the utterance indirectly through a paraphrase but might directly react to its communicative impact as a global message.[5]

A similar interpretational problem arises as far as the impact of the question formulated among close acquaintances in a Polish context: *Czy Pani sama uszyła tę sukienkę?* [Did you sow this dress by yourself?]. This utterance having a syntactic form of an interrogative sentence may be understood by a Polish lady in pragmatic terms either as a critical remark about her dress or as a compliment. Its interpretation depends upon the evaluation of the relationship between the sender and receiver. The sender must be very well known by the receiver as a sincere or envious malicious person. The propositional content included in the logical semantic structure of the utterance: "Did you sow this dress by yourself?" expecting alternatively a *Yes* or *No* answer, allows the receiver to draw at least two if not three conclusions. In the first case, it may mean: 'one can see that the dress was not sewn by a professional tailor'. In the second – 'the dress is absolutely perfect; you are a very gifted person, indeed'. The third interpretation may depend upon the intimacy between interlocutors

---

[4] Repetition, replacement, stressing of the spoken text, turn-taking rules and camouflage as exemplary kinds of nonverbal behavior are revealed by Jolanta Antas (2000: 213ff) on the basis of works on nonverbal communication by Mark L. Knapp (1978).

[5] Jerrold M. Sadock (1974) maintains, as Olga Sokołowska points it out "that in the case of some indirect speech acts the illocutionary force is semantic in character, and that it is encoded in the sentences accomplishing them from the very beginning of their derivation" (2001: 61).

and their common knowledge about the situation in the market, for example, 'what is done by hand is better then what the factory produces' or *vice versa*.

When we take into account two other, often cited, examples of the same kind, e.g.: *Why don't you sit down?* and/or: *Do you always make your own pastry at home?*, we may come to the conclusion that pragmatics is to be considered from the side of a sender who is the author of an utterance. It is, namely, the sender who endows his utterance with a certain pragmatic value.[6] The understanding of linguistic utterances in accordance with the intentions of a sender by their receiver depends, among the other things, upon the fact, whether the context (of the domain of reference), in which a given utterance appears, is the same for both the participants of communication.

3. Between philosophical and hard-science pragmatics

What scientists are able to observe in the pragmatics of linguistic communication is the verbal behavior of people and their relevant non-verbal behavior under the influence of verbal stimuli. From the viewpoint of physics, only the energy flow that comes into being as connections between the communicators may be evaluated as constituting a measurable phenomenon. When the communication takes place, there is always a certain amount of energy expended by the individuals who produce and receive intentional semantic stimuli. The content of intentional speaking, however, cannot be directly tested. It may be inferred through the intersubjective knowledge of communication participants.

What can be documented with certainty is the fact that communicating individuals unite into groups forming dynamic and steadily changing linguistic linkages of a collective character thanks to the interaction through the vocal-auditory channel. Considering the duration of these interacting groups, the question arises: to what extent do they exist as real

---

6   While "[c]riticising the Performative Theory of speech acts, which postulated the deleted-performative-clause explanation", Dennis W. Stampe, "proposes that more heed should be paid to the speaker's intention, instead of attempting to explain illocutionary force solely in terms of convention – linguistic or social", as quoted by Sokołowska (2001: 50).

entities describable in the role of communicative[7] or discursive communities, or are they only assumable as theoretical constructs?

In the domain of nonverbal behavior, the most elementary activities of the human body: the electrochemical activities, self-moving activities, connected with the biological nature of humankind, can be registered by testing, because they belong to the physical domain of investigation. It is undeniable, therefore, that the more unique properties of human nature such as the activities of feeling and thinking, are those connected with the self-awareness of communicating individuals, couldn't be performed without physiological activities and without the unconscious activities of bodily organs, as well as the consciously controlled movements of the hands, legs, head, in the process of interpersonal communication etc.[8] These remain, however, beyond the interest sphere of linguistics proper.

Most of the work conducted hitherto in the domain of linguistic pragmatics, which aims at understanding the nature of verbal communication, and the effectiveness of speech acts in dependence of environmental conditionings, is based on philosophical foundations. The focus of pragmalinguistic or sociopragmatic studies dealing with the issues of language communication is concentrated mainly around the search for aims, intentions or tasks of the participants of interpersonal communication. And what the communicators intend or have in view, what tasks they want to achieve are indeed those facts that cannot be observable directly. Unquestionably, the dispositional properties of communicating

---

7   It might be important to recall the views of Andrzej Gawroński or, *inter alia*, Norbert Reiter assuming that the language of collectivity does not exist or that the communicative community is a fiction (cf. Wąsik, Z. 2001b: 29).

8   The distinction of the four types of human activities is derived from the theory of communication where the human being is analyzed as the Self. It is worthwhile, therefore, to be familiar with the model of the Self as a "Semantic Reactor" adapted from J. Samuel Bois (1973: *The Art of Awareness: A Text on General Semantics and Epistemics*. 2nd ed., Dubuque, Iowa: Brown, p. 20) in DeVito 1976: 63ff. Referring to the studies on the determinants of the Self, one can notice that the feeling operations as, for example, needs and drives, wants and fears, hopes and ambitions, as well as love and hate, commitment and indifference, trust and distrust, happiness and sorrow, contentment and frustration, as well as thinking operations, for example, adding and subtracting, conceptualizing and abstracting, decision making and strategy formulations, and the like, are connected with the symbolic activities of senders and require interpretative activities of receivers. They depend, undoubtedly, on the electrochemical and self-moving activities of the Self and his awareness of both interacting individuals in the communicational context.

individuals can only be deduced from the introspective knowledge of receivers, who assume one another as having similar experiences as they impute to senders.

Examining the way how people exchange a few words, one can only state in terms of communication theory, the only fact which is observable is the ways people enter into interactions. However, as regards the transactions that occur between them, only the interlocutors, and no-one else, can deduce, or preview, their outcomes, or can elicit or adapt to changeable conditionings of interpersonal relationships.[9]

The communication participants we are interested in from a human-centered perspective actualize the relationships between the domain of expression and the domain of reference each time when they interact verbally. Nevertheless, the only observable thing are the links through sound waves exchanged between the speaker and the listener and not how they interpret the meaning bearers, called sign-vehicles in semiotics.

Similarly, while appealing to sign-and-meaning-related terms, one may observe the extra-semiotic reality to which the sign is referred by its user, but the referential value of the sign-vehicle is to be deduced from the linguistic or social and cultural context. Therefore, it is important to distinguish between the observable and the concluded reality of the domain of reference, called in terms of linguistics also extra-linguistic reality.

It seems obvious that in the investigative field of human-centered linguistics, scientists are not in a position to study the linguistic proprieties of people solely from a hard-science perspective. The logical domain appears to be indispensable as a counterpart of the physical domain. In this context, it might be relevant to recall the ideas of the German philosopher Ernst Cassirer (1944) who argued that the nature of man cannot be discovered in the same way in which we approach the nature of physical objects. This means, that only physical objects can be described in terms of their objective properties. Man, in Cassirer's view (1944: 41),

---

9   Even Victor H. Yngve, though a trained physicist, deriving his conceptions from a hard science perspective is convinced that the interpretations of linguistic utterances are in the heads of the speakers and hearers. Cf. his stipulations (Yngve 1996: 13): "Is there any merit to the view that in nature we find only the physical sound waves, their interpretation being in the heads of the speakers and hearers? And does it not then follow that a scientific analysis must include the study of physical sounds themselves and a careful and detailed study of the people who produce and interpret the sounds? I think there is merit to this suggestion, but it needs further study."

can be described and defined only in terms of his consciousness. As Cassirer also claimed, only immediate contacts with people enables us to gain insight into the characteristic properties of man. In this, Cassirer showed his adherence to Socrates, the Athenian philosopher, thanks to whom, from early antiquity, philosophical reasoning switched from cosmological thought to anthropological thought.

In turn, one should also mention Adam Schaff, the Polish philosopher of critical-Marxist orientation, who expressed his conviction that we are not able to cognize human nature unless we approach it through dialogical or dialectic reflection (cf. Schaff 126–127). Above all, Schaff spoke against any attempts to change the philosophical interpretation of the reality of humans to the manner of the so-called "exact" or "hard" sciences. In particular, Schaff criticized the neo-positivist hypothesis, which assumed the unification of science through their reduction to physics as advocated by Rudolf Carnap, Otto Neurath and Moritz Schlick, and according to which there is no difference between the natural and psychical domains of subjects as organisms. Following Schaff's view we have to reject physicalism, which postulates that every scientific statement, other than the necessary statements of logic and mathematics, is to be translated into the language related to physical bodies, for example the statements from psychology into statements pertaining to the state of organisms (worth consulting is here the dictionary of terms and philosophical notions elaborated by Podsiad 2000).

As one may gather from Schaff's conclusions, the neo-positivist approaches postulated by the representatives of the Vienna Circle – concerned with positive facts and phenomena while excluding speculation upon ultimate causes or origin – assumed, in the end, the form of a simplified behaviorism. Behaviorists believed that human psychology just as animal psychology can be accurately explained through the measurement and analysis of objectively observable and quantifiable behavioral events, in contrast with subjective mental states.

The followers or opponents of the two different ways of thinking, physicalism or mentalism, as ascribed to Galileo[10] and Descartes[11] whose ideas played an important role in the formation of the philosophy of modern science, should bear in mind the opinions about their heritage today.

---

10 Galileo Galilei, Italian physicist and astronomer
11 René Descartes, known as Cartesius, French philosopher and mathematician

Even though both Galileo and Descartes represented initially the same conviction that the whole universe is composed of a uniform matter, which underlies universal laws of physics, "the followers of these two found themselves parting ways", as John N. Deely (2000: 15–16), the contemporary American philosopher working in the domain of semiotics, has rightly stated.

On the one hand is situated "the line of Galileans leading to Newton, Einstein, and Mission Control in Houston and placing men on the moon and ships bound for the stars",[12] and on the other "the line of Cartesians leading to Hume and Kant and a reluctant conviction that the universe of reality prejacent to and independent of the human mind is forever unknowable".[13] Instigated by Cartesian dualism which asserts that the thinking substance is independent of the universe of matter, the latter line contributed in the philosophy of man to the speculative understanding of the world. Even so, one can agree, in the end, explorations in the body-and-mind related pragmatics of human communication have enormously enriched our knowledge concerning the properties of communicating individuals and groups studied in the domain of contemporary psychology and sociology.

---

12 Sir Isaac Newton, English physicist and mathematician; Albert Einstein, German physicist, and U.S. citizen from 1940
13 David Hume, Scottish philosopher and historian; Immanuel Kant, German philosopher

Conclusions

In favor of the notion of grammar as a network of interpersonal and intersubjective linkages

The following treatise recapitulates the achievements of the author's inquiry into the nature of language as a tool or a property of man entering the domain of metalinguistic relationships between the sciences of language and the sciences of man. Referring the notion of the tool to functional objects which play a serviceable role for the sake of man, it has been argued, in chapter II, that the discussion on the functions of language, exposed by the practitioners of language and communication sciences, should depart from the analysis of the term *function* and the account of methodological consequences of functionalism as an interdisciplinary investigative perspective. It has been observed at the same time that in various investigative domains of scientific disciplines the notion of function was understood not in the same manner in dependence of the position of the functional objects in question, their inherent and relational properties belonging to different ontological categories in nature and culture.

Reflections on functions were made in the contexts of ascribing a serviceable role to various aspects and constituents of things, for example, to artifacts or parts of artifacts, organisms in general or organs as parts of organisms, as well as to the activities of artifacts or organisms and their parts, events or relations, and the like. Subsequently, it has been underlined that the functional thinking is based in practice on the substantiation of the existence of a given element in relation to its utilitarian property or on the search for a role which a given element fulfills in a given setting or for a given surrounding.

It has been also pointed out to the importance which the so-called etiological approach had for the formation of the notion of function. The distinction between etiology and teleology proposed by philosophers of nature, appeared to be crucial in confronting the cause-and-effect- to purpose- or goal-oriented way or reasoning. It meant for a scientist or observer to be free from handling the notion of functions as connected with aims-in-view which depend upon the dispositional properties of a conscious agent or from internally driven mechanisms of a goal-oriented system put into motion by an undetermined external causer.

Against the background of the ambiguity of teleological (purpose-or-goal-oriented) or etiological (cause-and-effect-oriented) understanding of the term *function* pertaining to intended or not intended ends resulting from its fulfillment, the author of this work proposed to distinguish between instrument-related and organism-related functionalism with reference to the functional outlook on the nature of language. From the perspective of functionalist instrumentalism the fulfillment of a function or functions by a given element was seen as depending on the external purpose-or-goal-oriented stance of acting subjects or from the cause-and-effect-oriented conditionings of objects underlying those respective acts.

The focal point of organism-related functionalism, called here organicism, was the functional explanation elaborated in the philosophy of biology and biologistically-minded cultural anthropology searching for the goal or cause of existence or activity of a certain element in a determined setting of a living system.

Discussing the reception of instrumentalist functionalism in linguistics and other related disciplines, it has been stated that the researchers who treated language in general and linguistics sign in particular as a tool used by people for describing extra-linguistic reality or as a tool for communicating purposes had spoken initially about one principal function of language. They took as a core of interest the purpose-oriented rationality principle developed in urban architecture and the applied arts. According to this principle producers of utility goods were interested in revealing such properties in their structure that might accommodate them functionally to the needs of their utilizers. Based on the principle of abstracting relevance their functional approach to the object of investigation had led to the detachment of functionally relevant from functionally irrelevant properties, which was one of the main basis of detaching phonology from phonetics in the Prague school linguistics.

A critical survey of subsequent steps in the development of functional thinking in the sciences of language allows the reader to notice the impact of organicist functionalism, derived from biology and continued in the behavioral and social sciences, where the fulfillment of a function by a given element was treated as depending on the homeostasis of a given living system as a self-sufficient being. It has been observed that numerous searches for the functions of language were reduced overwhelmingly to the listing of tasks, aims-in-view, purposes or intentions of communication participants attained due to the serviceable role of verbal means. From such a view the essence of functionalism had been in fact lost as regards the search for the functionally relevant properties of meaning bearers realized in linguistic texts. The same might also be said about the hierarchy of their components. With regard to language as a whole it has been revealed that all hitherto distinguished functions belong to one and the same communicational function of verbal meaning bearers and their extra-linguistics objects of reference. Subordination and superordination of such functions realized by linguistic utterances depend upon the hierarchy of interpersonal tasks realized through social communication. Put more precisely it may be said that the attention was paid to the fact that communicating individuals taking part in determined domains of social life realize various needs, purposes and aims-in-view or tasks through conventionalized practices of verbal behavior, being possible due to linguistic structures fossilized in their mental spaces as a result of steadily repeated forms of interactions. Hence, the inquiries into the functional nature of language conducted in the domain of sociology, allow the researchers to treat the phenomenon of language as a whole as both the physical and psychical property of language acquired by individuals in the social environment.

It has been pointed out that linguistic communication contributes to forming group communities of those participants who have something in common. Such commonalities are formed on the basis of collectively agreed meaning bearers, widely accepted and understood, due to common references to the same fragment of the extra-linguistic reality on the basis of the same communicative values, to which communication participants strive. It should be remembered that the meaning in communicational acts is in each case relativized by communicating subjects depending on the fact how they refer the meaning bearers to the extra-linguistic reality, agreeing the scope of their understanding. It has been observed also that the focus on verbal communication from the viewpoint of human-centered linguistics, gives the possibility of paying attention to the multiple relationships between linguistic properties of people both in their natural and cultural environments, as well as particular interactions between various members of linguistic communities which find themselves in determined contact situations. The aim of such approaches is to search for pragmatic correlates of human communication contributing to the differentiation of linguistic communities on various levels of social stratification in various steps of development. It is assumed therefore that linguists working in the domain of human sciences have at their disposal a broad assortment of factors, cultural, social, political, economic and technological, which have an impact upon the dynamics of changes in the linguistic properties of communication participants, characteristic for their nature as human beings, persons and members of commutes playing certain roles in the local and global scale.

On the basis of a human-centered approach the point of departure of the thesis put forward in this work is that language does not exist as an autonomous phenomenon in itself, as far as it is connected in its genesis and functioning with the communicational activity of man. In fact, the only accessible objects to external observations are communicating individuals. Observable also are changes in communicating individuals, when they interact with other individuals, thus forming interpersonal and intersubjective linkages. Hence, the focus of a linguist's interest sphere should encompass those poperties of communicating individuals that unite them as members of group linkages when they interact communicatively with their social, cultural and natural environments while being engaged in the realization of common communicational tasks. In consequence, language is regarded here not as an external tool of man but as as his relational property, as far as its existence may be deduced, firstly, from observable interactions based on the utlization of common means and, secondly, from the concluded attainment of correspondences that might exists in the task-oriented realization of mutual understanding processes. What the researcher of human-centered linguistics can, therefore, state is an obvious fact that people unite themselves into collectivities of different temporal and spatial character on the basis of two types of communicational linkages which might be both concretely observed and assumed mentally.

The first type of linkages is accessible to the external observation of hard scientists studying the communicating individuals when they are interpersonally united by sound waves or their surrogates (transformed into sound waves in the process of their mental reception) playing the role of verbal meaning bearers. In such linkage types, where people, as communication participants, are engaged in sending and receiving messages, relevant is the amount of energy expended by individuals which might be measured in terms of acoustic physics. The second type of linkages appears in the formation of changeable linguistic knowledge about the concluded reality when communicating individuals enter into intersubjective relationships on the basis of commonalities of their experience in the cognition and interperetation of verbal messages as having the same or similar referential value. As such, the study of this second type of linkages enters into the domain of logical inferences utlizing only the self-referential introspection as a tool of reasoning.

The basic assumption of this treatise is the statement that the study of language in relation to its genesis and the ways of functioning should be concentrated on the search for the linguistic properties of man materialized in the products and acts of communication. Departing from a human-centered perspective, it assesses in detail those investigative standpoints in which man and his communicational skills are seen as the primary object of linguistic studies. Exposed are mainly issues pertaining to the distinction of the linguistic properties of people in the context of language acquisition and language contacts between native language bearers and foreign language users.

The linguistic properties of man are defined, according to the state of knowledge in human-centered linguistics (called human linguistics in Victor H. Yngve's proposal), or homolinguistics for short, and applied linguistics, in terms of knowledge, aptitudes, skills, habits and activities of communicating individuals that are shaped on the basis of their participation in communicational events. They allow the people, as speakers or hearers, to produce and to interpret, to distinguish and delimit utterances with regard to their communicative values, as, e.g., indexical, signaling, semantic and pragmatic. In the enumeration of factors conditioning an effective communication in speech and writing, attention is paid to the capabilities of individuals that are indispensable for their functioning in the various domains of social life of an interpersonal, public or mass character. Taking into account that the contacts among people in individual communicational practices are realized through socially accepted patterns of verbal conduct, the author of this work proposes to distinguish between the realizational and dispositional properties of man. The exposure of the linguistic properties of people which develop communicative linkages on the basis of observational and concluded similarities in their referential and interpretational activities, points to the existence of language not only in concrete interactions but also in mental intersubjective knowledge. These linkages manifest themselves in the communicational acts when people understand each other while referring the meaning bearers to the extra-linguistic reality, and interpreting them in a similar way depending on the ecological environments in which they live.

In exposing the "ecosystemic" nature of the linguistic properties of communicating individuals, the author postulates to investigate the social network of semiotic-communicational linkages, in which those individuals function in the capacity of communication participants, as a kind of ecological grammar. In reality, this "grammar" is dynamic when taking into account the fact that the linguistic properties of persons participating in social linkages, constantly change under the influence of ecological conditionings (biological, psychical, social, cultural, technical, political, economic, and others, cf. page 85), which can shape and determine the modes of their formation into autonomous entities constituting separate interactional and communicational groupings. One could examine the possible expansion of such linkages interacting with each other in time and space along with their particular, ethnic, national, confessional, professional or cultural-natural ecosystems. Working in the field of human sciences, a linguist might be interested in exploring a wide range of ecological factors that affect the changes of the linguistic properties of communicating individuals intrinsic to their dispositional qualities as representatives of biological species and their role-manifested features as communication participants on a global scale. These studies might exhibit certain external factors that sustain or threaten the vitality and maintenance of the hitherto-existing linguistic communities in their local environments.

The "ecological grammar", accepted by the author of this work, excludes the rationalist hypothesis that linguists may be able to deduce from the languages of the world elements and structures that are primordial to human thinking. In contradiction to "universal grammar", the ecological grammar is considered as growing out of the experience of practitioners of human sciences who have noticed that the manifestation forms of verbal behavior of communicating individuals and communicative linkages become polymorphous while being unequally realized and patterned under the influence of their environmental conditionings.

Entering the domain of "ecological grammar", the author postulates to study communicating individuals when they play their roles as participants of social interactions, in abstraction from the fact that they, as persons, are also organisms. In such a human-centered perspective, the concept of the ecology of verbal discourse is encompassing the interrelationships between the linguistic properties of communicating individuals and groups within the surroundings in which they function, and to which they refer their expressions.

The application of the perspective of human-centered linguistics allows the researcher to treat communicative linkages (monolingual, bilingual or multilingual) in a dynamic way. Heretofore, the observer's knowledge of the communicational processes within social groups had to be supplemented by considering changes, because communicating individuals as persons participating in social interactions always change and adapt to progressive changes in ecological conditionings, in which centers and peripheries, densities and diffusions may occur as far as the distribution of linguistic properties is concerned.

Accepted in this framework, the notion of the linguistic ecology refers to verbal interactions between communicating individuals and social groups that unite themselves into linkages and linguistic communities with their environments, in which they perform discursive practices and acquire discursive patterns, while interpreting and referring new utterances and expressions as meaning bearers to the steadily changing extra-linguistic reality. Exposing the biological conception of the linguistic ecosystem, the center of which constitutes the communicating individual as a participant of group communication bound with other participants through understanding in a given language, one assumes that the basic ecologically determined linkage is a group that consists of at least of two people. Having in mind the significance of verbal communication in the formation of eco-linguistic communities, the author has elaborated the conception of an ecological grammar of interpersonal and intersubjective linkages that exposes the functioning of verbal means produced and understood by communicating individuals on various levels of social groupings. In a human-centered perspective, ecological grammar is to be discovered in the networks of relationships between communication participants who are united by verbal means on the one hand in changeable practices of their sending and receiving and, on the other, in stabilizing patterns of their interpreting and understanding. Each communicating individual possesses such linguistically relevant properties that are observable, and that might be assumed on the basis of inferences. Communicating individuals as participants of communicational events also become constitutive members of linguistic linkages of various types and larger collectivities forming aggregations of task-oriented linkages. In the network of such collective interactions individual properties are manifested in the practices of message production and their transmission but when received they become typified as patterns. The greater the network of contacts, the more possibility that such individual properties become generalized and widespread. As opposed to the traditional view of speech communities linguistic linkages have a dynamic character; they do not form any uniform systems. In ecological grammar, one can distinguish centers and peripheries with a dense core and blurred boundaries of distributed linguistic properties of individuals and groups. Through a sequence of repeated interactions, temporary linkage may became long-lasting. Patterns of behavior connected with the dispositional properties of people, internalized mentally, may become changed when they are realized in changeable communicational practice.

Reconstituted in the light of human-centered linguistics, some traditional views about language irrespective of their place in particular holistic models appear to be suitable and relevant for this theory in question. For example, language defined with regard to its abstract and social character is inapproachable, but only its concrete and individual realizations in verbal meaning bearers, called linguistic signs, which are referred by communicating agents to things and states of affairs lying in extra-linguistic reality due to social conventions, may be observed as mediating between their senders and receivers. Language belongs to the property of a particular human being who functions as a member of communicative linkages and linguistic communities, but it forms itself in the human activities of cognizing and utilizing verbal signs, which underlie constant interpretations, as far as their meaning depends on subjective ascriptions. While investigating the role of language in interpersonal communication the attention of external observers is drawn to the problem of self-awareness and the subjective intentionality of communicators. With regard to the pragmatic nature of interpersonal communication, one notices that, due to the typification processes of communicative behavior, all linguistic locutions resulting from their frequent occurrence become standardized as social constructions and assume over time an idiomatic character. This means in fact that language does not comprise a set of established fossilized labels.

To reveal the properties of language in relation to man, one postulates to accept the following set of at least five epistemological perspectives, which are decisive for the defining scope of the object linguistic studies, as: (1) collective solipsism, assuming that language constitutes common property of individual "selves" united into homogeneous communities which form a set of collectivities and not collections, (2) communicational pragmatism, departing from the belief that language functions while forming temporary linguistic linkages and long-lasting communities through the realization of the communicational tasks of individuals in respective speech act, (3) social constructivism, claiming that linguistic communities create apart from shared meaning bearers also an intersubjective world of meanings based on an inferential knowledge about reality to which they refer those meaning bearers for the purpose of mutual understanding, (4) "enactive" ecologism, which maintains that what people are talking about must be first cognized in reality through sensorial stimuli acting on their senses, (5) experiential "recentivism", which takes for granted that what is communicatively relevant for an individual, with regard to his actual and prospective acts of comprehension and interpretation, are solely his previous experiences with an outer and inner world recalled from the memory in the recent time.

Against the background of a poststructuralist attitude where language has been studied in relation to man, one might perhaps expect that more and more descriptions of personality traits of individuals interacting in changeable biological, cultural and social conditionings will be taken into consideration as communication variables, for example, values, needs and motivations, as well as expectations and requirements.

Supplement One

Applying an ecological model of language to the external characteristics of Frisian

The subject matter of this chapter is to illustrate how the theoretical model of the ecology of language, elaborated and presented at various international conferences (Wąsik, E. 1998: 54–55, and 1999c: 57–64), has been utilized for the description of Frisian as one of the minority languages functioning in Europe. It includes the criteria and methodological positions of those linguists who postulated to detach the ecological properties of language from the systemic properties of language (Zawadowski 1966) and those who dealt in their material studies with the external characteristics of minority languages from the perspective of external linguistics (Wąsik, Z. 1993a). The constituents of this model embrace those ecological-relational variables of natural languages which have an impact upon their formation and functioning in the environment of individuals, communities, societies and nations.

Ecological variables (as distinguished in Table 1, chapter IV), have been grouped here around such issues as: (I) the position of a language in all hitherto existing classifications and its relation to other languages and communicational semiotic systems, i.e., its metalinguistic ecology, (II) the descriptive ecology of language bearers and language users, as well as (III) the characteristics of communication levels connected with various types of texts, language varieties, domains, functions and situations of their use and the like, i.e., the ecology of language communication. Having elaborated a detailed typological matrix for the description of minority languages, the author of this chapter evaluates the ecological-linguistic status of Frisian as a set of dialectal vernaculars from the viewpoint of: (1) its place in genetic and typological classifications, (2) the etymology of its name and the name of its users, (3) the social stratification of its users and (4) their territorial, geographical and political settings, (5) its external history and the history of its bearers. Further constituents of the following matrix of ecological variables include also: (6) human attitudes towards a language – language as a criterion of ethnic identity and other semiotic systems characterizing language bearers; language loyalty and ethnic solidarity, (7) language policy and language planning – standardization, codification, and the maintenance of the communicational autonomy and vitality of a language; organizational and political support for a language and forms of struggle for its maintenance, cultivation and education, (8) the media-related realization of a language – language varieties, (9) domains and functions of language use, (10) the status and intimacy of a language in symbiosis or conflict with other languages in contact situations.

The external characteristics of Frisian in accordance with the criteria elaborated in the following model of an ecological description of languages has enabled the author to draw two kinds of conclusions pertaining to its ecological situation as a minority language and to evaluate the state of the art in the domain of Frisian studies.

1. The ecological status of Frisian in the Netherlands and the Federal Republic of Germany

(1) Frisian, as a set of regional dialects and local vernaculars, belongs along with English, Dutch and German to the West-Germanic branch of the northern group of languages. Their linguistic substrate is not known up till now. One believes, however, that the users of Frisian must have communicated at the beginning in a non-Germanic language, i.e., supposedly in a kind of linguistic substrate which was then superposed by the language of Teutons during the Roman rule.

In the historical typology of Teutonic languages the language of Frisians belongs to the same linguistic type according to its internal structure (as far as its standard variety spoken in Friesland, the province of Netherlands, is concerned) as, for example, Modern Dutch. It finds itself at the same stage of development as adjacent languages bordering with it geographically more or less until the 16th century. In the geographic-cultural classifications of the languages of Europe, Frisian located itself within the same littoral Sprachbund as the languages of those countries, the existence of which was connected with the access to the North Sea.

(2) The tribal name of the users of Frisian and that of the language itself is considered, in the light of contemporary estimations, as non-

Germanic (i.e., neither Celtic nor Teutonic but rather Proto-Germanic). The name of the language of Frisians, called by their native bearers in the Netherlands as *Frysk*, and in Schleswig, accordingly, *friisk*, *freesk*, *fråsch*, *freesch*, is also to be encountered in some regions as a synonym of local vernaculars. As such, therefore, the present name *Frisian* (or also the name determining the Frisianness of a given linguistic community) is misleading, especially when it is referred to the Low German varieties spoken on the territory of the former county Friesland from the times when the Frisian language had been wiped out by the dialects of German which started to dominate there. Furthermore some users of the local varieties of Frisian have preferred to determine their means of communication by geographical names derived from the regions in which they live. For example, East Frisian spoken in Saterland is known as Seeltersk, and the Frisian dialects spoken on the North-Frisian Islands are identified according to the name of the island where they are used: Sölring, Fering, Öömrang and Halunder.

(3) Following the criteria of social hierarchy, the speakers of Frisian who inhabit the province of Friesland come mainly from the poor countryside, and at present they represent the most numerous group forming the lowest social stratum, with the lowest education and the lowest income. By contrast, in the North Friesland, the place of Frisian speakers in the social ladder is determined by the place where they live (on the island Sylt, for example, the Frisian-speaking inhabitants belong to the wealthy middle class, whereas on the continent to the lowest class). The age of the users of Frisian coincides with the degree of how far it has been acquired (the representatives of elder generations have achieved greater command of the capacity of writing and reading in their native language); however, no differences are observed in the degree of language acquisition among men and women communicating in Frisian as their native vernacular.

(4) Frisians form an autochthonous ethnic minority which is politically recognized in the Kingdom of Netherlands and the Federal Republic of Germany. They live dispersed in the Province of Friesland and on the West Frisian Islands (around 400 thousand). Small-group communities may be encountered in Saterland (approximately two thousand), in Schleswig and Hanover, and on the North-Frisian Islands (around 10 thousand). As it results from the dialectal and geographical classification of Frisians, known as an ethnic minority living in dispersion mainly in

the territories of the Kingdom of Netherlands and the Federal Republic of Germany, they are divided into three main branches, which differ also with respect to their number, i.e., Middle Frisians (or West Frisians), North Frisians and East Frisians.

(5) With reference to the present and past estimation of linguistic facts, Frisians have been included into the groups of Teutonic tribes. But as to the proof of some historical and archeological data, supported also by some linguistic conclusions based on onomastics, there is a hypothesis that Frisians descend from a third (apart from Teutonic and Celtic) Indo-European group of peoples, called Nordwestblock. They have constituted the so-called *Verkehrsgemeinschaft* (using the German term approprietly) based on etymological principles, without having any form of statehood, but being always united in the struggle with the sea and thanks to the possession of common law.

Due to an ideology of freedom shared by all members of this ethnic group, which results from the continuation of legal fundamentals, Frisians have been able to keep their own national identity throughhout the ages; one can speak here about the linguistic separateness of Frisians starting from the 8th century. Nowadays, however, the most important symbol of Frisianness for Frisian speakers is their native language. Frisians do not constitute at present a compact group neither from the political nor from religious point of view.

(6) The results of linguistic inquiry into the position the people prefer to take with regard to their native means of communication have shown that the attitudes of the users of Frisian toward their language are positive; however the pro-Frisian language attitude is rarely met in the case when the social status of the users of Frisian becomes higher. In the Dutch province of Friesland, the ethnicity of Frisian-speaking individuals is estimated according to the linguistic criterion of whether one is a native user of Frisian or not.

The attitudes toward North-Frisian dialects and the degree of the linguistic and ethnic consciousness of Frisians living in North Friesland are not identical in different parts of the region. Linguistic data have shown that the ethnic community of Risum-Lindholm constitutes there the mainstay of Frisianness. Nevertheless, the internal stigmatization of the North Frisian dialects spoken on that particular region manifests itself in a negative attitude toward vernaculars other than their own Frisian one.

(7) Frisians strive for equal rights for their own language as a rule by using peaceful and legal means. In the province of Friesland, organizations cultivating the Frisian language and culture started to be created at the beginning of the 19th century. As a result, at the present time Frisian has a strong institutional support being legally recognized as an official regional language. In North Friesland, there are approximately 20 unions concerned with language policy making, and the typical activity aimed at the maintenance of particular North-Frisian, starting from the second half of the 18th century, has been until now lexicography. Frisians in Schleswig posses a formally assured right to support and to cultivate their own native language, but legally they are not recognized as a second national minority (along with Danish) on the territory of the Schleswig-Holstein state.

(8) The analysis of historical documents has shown that the texts written in Old-Frisian (among which the largest part constitutes legal texts having the character of private notes used as handbooks of old customary law) occupy a special place in the Germanic cultural tradition both with respect to language and to content. On the basis of these texts it is known, among the other things, that the golden age in the development of Frisian took place since the 13th century when the Old-Frisian legal texts were written in the East Frisian and later West Frisian territories.

(9) At this point in time the use of Frisian is limited to dyadic and small group communications, for the most part at home, among neighbors or in primary education. Actually, Frisians in the Netherlands are for the most cases, out of necessity bilingual, using Dutch as a superior language with regard to Frisian, and in the Federal Republic of Germany bi- or trilingual where High German and sometimes also Low German occur together with Frisian.

The linguistic situation among Frisian communities is to be characterized in terms of a *diglossia* state where an alternative use of the native language or the local dialect or the language dominating in the whole state occurs. The state of an ideal balance does not exist at all because languages spoken, as a means of communication, by native speakers cannot achieve the same usefulness in all domains of social life. In the course of the last few years, it has become widespread also in the court of justice, in the regional parliament, and in the church. Its use has been declining in some areas, but through different organizational forms it now embraces several new domains of social life.

(10) The established position of Frisian as an official regional language has contributed to the fact that the consciousness of ethnic solidarity among Frisians is steadily growing. Moreover, the feeling of national pride for their belonging to a distinct language community usually increases the loyalty of individual speakers toward their own language. At the same time, Frisians are aware of the situations and domains of communication in which their language has to give way to other languages serving as a means of national or international communication. To come to the point, one can ascertain that the full characteristics of Frisian might be possible when the practitioners of language sciences are enabled to utilize also comparative studies on the ecology of other languages of the world.

2. The state of the art in ecological studies on Frisian

To the next group of conclusions belong statements based on the state-of-the-art in the Frisian studies. On the basis of descriptive data one can observe that the practitioners of language sciences studying Frisian from the viewpoint of their external properties have utilized the theoretical achievements both of the internal and external linguistics.

- The language of Frisians is not to the same degree investigated and described, when one takes into consideration its division into three basic dialectal groups, West Frisian, East Frisian and North Frisian. It seems that the external facts of the Frisian language, – namely those resulting from purely linguistic statements (deduced from its systemic properties and from the properties of its particular dialects) allow us to specify its place in current linguistic classifications or to distinguish its varieties (especially regional) both in the past and in the present time – were described by Frisian linguists in a more exhaustive way. For researchers working in the domain of Frisian studies linguistics constituted an instrument of the Frisian Movement in the province of Friesland in the Netherlands. Therefore, their most important task was to collect and to elaborate historical, linguistic and literary facts from the Frisian past that were meant to strengthen the national identity of native speakers of Frisian. The objects of material studies were also vernaculars spoken by Frisians (hence the idea to develop the Frisian dialectology). In the domain of synchronic studies, the Frisian language as a whole has not been yet described from a typological and contrastive perspective as far as its systemic properties are concerned. Its phonological, morphological, syntactic and lexical peculiarities were not pursued independently of the linguistic approaches in terms of kinship and genealogical descent which would permit researchers to determine the place of Frisian in the structural typologies of languages in general and in the typology of language structures in particular. This statement refers mainly to its systemic-functional characteristics in comparison to other non-Germanic Indo-European languages. Although such typological studies were carried out in the domain of North-Frisian dialects, their task was to determine their place among the other Teutonic dialects spoken in the northern regions.

- Starting from the 1960s Frisian was studied from a sociological perspective, and the representatives of other non-linguistic disciplines, were also active in contributing to the knowledge of the external conditionings of the language users. Works providing theoretical foundations for interdisciplinary studies on the language in general, which were published in the 1960s and 1970s, were utilized as sources of inspiration in studies of vernaculars spoken by Frisians particularly in the province of Friesland. But they were not applied to the greater extent with reference to the studies on Frisian spoken in Schleswig or in Saterland.

- The three principal territorial varieties of Frisian, i.e.,West Frisian, East Frisian, and North Frisian, as far as their environmental dependencies are concerned, evoked the interest of researchers to an unequal degree. The diversity of interest is to be explained, *inter alia,* through the different degrees of language consciousness and through the strength of respective speech communities expressed in their demographic quota and, respectively, also in the scientific activity of their indigenous members striving towards the mastering and cultivation of their own native language. As far as the external conditions of Frisian is concerned, there are no comparable statistical data coming from studies which deal with the description of the dialects spoken in all the three Frisian territories, conducted from the same investigative perspectives. Deductive-empirical research conducted in the three regions where Frisian is spoken was based on various criteria, which was in fact incommensurable, and disproportionate with regard to the scope of investigative domains.

- The scientists who conduct research on New West Frisian and its socio-political and cultural environment have utilized current models and theoretical patterns of study. Moreover, some modern methods and investigative techniques elaborated by sociolinguists (or the sociologists of language), which constitute mostly widespread frameworks for examining the problems of societal bilingualism or multilingualism have found also its application mainly in the province of Friesland.

- Regarding the characteristics of the external situation of North-Frisian dialects, as well as of the Saterland vernacular, the practitioners of language sciences refer for the most part to the results of historical studies, when taking into account the factors codetermining the regress of North- and East Frisian.

- The description of Frisian as a whole, i.e., in its entirety, pertaining to all aspects of its situation, domains of occurrence, functions and social stratification, etc. has not been the subject-matter of major works. What we can encounter as hitherto prevailing are mostly articles and books of historical-ordering character, representative for particular areas of respective studies, e.g., reports on Frisian language and culture in the Dutch province of Friesland and the same can be said also about the Frisian-speaking part of Schleswig-Holstein Bundesland. From our personal viewpoint, we may remark that the current studies on the Frisian language, carried out particularly in the Frisian Academy in the domain of the sociology of language, seem to show an acceptance of the ecological situation of Frisian as a minority language. Research workers continue to concentrate on such issues as, e.g., the use of Frisian by the administration, the patterns of verbal behavior in interpersonal relationships, and the degree of the acquisition of Frisian as a second language by the non-Frisian language speakers.

Supplement Two

Exposing the markers of Frisian ethnicity through a semiotic perspective

The subject matter of this part (based on the author's paper, see Wąsik, E. 2001) encompasses the semiotic exponents of ethnicity applied to Frisians. The term *ethnicity*, etymologically connected with the Greek word *éthnos*, has been borrowed from the sociological discourse for describing a collective of self-conscious people united or closely related by shared experiences. Among factors subordinated under the concept of ethnic identity, one may enumerate, e.g., the name of language bearers and its users, the myth of common descent, the uniquenes of culture, the association of belonging with a specific territory, and the sense of solidarity (cf., e.g., Erickson 1987: 91; Devetak 1996: 203–204).

To illustrate the mechanism of individual and collective expression of ethnic identity we have selected West Frisians, one of the three main branches of Frisians – along with North Frisians and East Frisians – living in the Kingdom of the Netherlands and the Federal Republic of Germany. They form a compact ethnic and linguistic minority in the Dutch province of Friesland. The language of all Frisians belongs to the family of West Germanic languages (similarly as English, Dutch and German). Despite their genetic affinity with the North Frisians or East Frisians in Germany, the West Frisians, can be regarded as different from a cultural and linguistic point of view. In fact, West Frisian constitutes a group of dialects, which vary in a significant degree from other Frisian dialects. The source material for our contribution (referring to the monograph of Wąsik, E. 1999b) comes from the empirical investigations conducted by Frisian scientists, mainly in the field of literary studies and sociology.

Following the demographic data of the 1990s provided by Gorter (1997a: 1152), the number of West Frisians amounts to 600 000. This means that they form less than 4% of the total population of the Netherlands. It is remarkable – in view of their history and language – that the West Frisians do not strive for their own statehood, but rather they only want to maintain their unique position within the broader frame of Dutch culture. Nevertheless, they are aware of their situation in comparison to other ethnic groups in the European Union (cf., e.g., Breuker, Janse 1997: 10–11).

Among the facts which affirm the distinctiveness of Frisians, special attention deserve the scientific hypotheses about the origin and division of this people mentioned in history first by Pliny in 12 B.C.[1] One of them, for instance, a hypothesis coming from the 60s, has suggested that Frisians descend from a third (apart from Teutonic and Celtic) Indo-European group of peoples, called *Nordwestblock*.

In the earliest past, Frisians must have constituted a speech community on the basis of etymological principles. After all, it is not easy to understand why and in what way they have managed to keep their ethnic identity through ages. The distinctness of Frisians is often explained through the existence of the Frisian ideology, common for members of this ethnic group.

Starting from the Middle Ages, Frisians have striven to keep their political independence because of the alleged privileges granted to them by Charles the Great. Presumably, he ordered in the year 802 to translate the Frisian legal law (*Lex Frisionum*) into Latin. This assumption seems, however, to be problematic, because the oldest of the preserved Old Frisian texts, mainly legal sources, comes from the 13th and 14th centuries. Apart from the law, another factor uniting all Frisians in the past may be also enumerated as their common task, namely, the struggle with the sea.

The topic "Frisian freedom" is regarded as a crucial motive in the history of Frisian ethnicity. From the remotest times, Frisians have been associated with freedom. Their name has been confronted with the adjective *free* and even until today (as it is stated by Breuker, Janse 1997: 29–30) 'Frisian and free' (Du. *Fries en vrij*) is – at least for the majority of Dutch citizens – an obvious linkage of words.

---

1   Pliny the Elder, Gaius Plinius Secundus, Roman naturalist and writer. Cf. Seebold 2001: 479–481.

1. Myths and stories of the origin of Frisians as a reference source of ethnic separateness

The Frisians, as an ethnic group, have always shared the belief of common ancestry and common past. The myths of the origins of Frisians emerged as early as at the beginning of the Middle Ages. From the first decades of 16th century, special chroniclers were employed for the aim to describe the distant past of Frisians. A great interest in the origins of Frisians arose mainly around the year 1600, but the stories had been popular until the 19th century (Discussion about the myths and legends of Frisians are to be found in the works of Breuker, Janse 1997: 20ff).

Evaluations of literature from the Middle Ages (cf. Breuker and Janse 1997: 20–21) have shown that Frisians were seen at that time as disobedient and crude people. Besides, they were also associated with wealth, strength and exoticness, what probably – according to the opinion of Breuker, Janse – had its sources in *Frisia Magna* from the Merovingian and Carolingian period (from A.D. 476 until 751 and from 751 until 911).

If we ponder the value of myths that deal with the origins of Frisians, we had to refer to the well-known legends summarized by Breuker, Janse (1997: 23ff). Such texts, as *Als wy lesset in cronicis* – a short rimed chronicle from the 15th century, *Coronike van Vrieslant* from the second half of the 15th century and *Historia Frisiae* from the end of the 15th century would constitute the source of our reference. Furthermore, the works of the authors from the Renaissance period should also be mentioned here.

The medieval works include various versions of an old legend of how travelers arriving by sea from Asia to the northern countries became founders of new peoples (or new nations). The story had been modified by the introduction of a certain Frisian personality, whose name was *Friso*. Apart from the daughter, he had also seven sons, and each of them he gave to rule one part of *Frisia*.

As it results from *Historia Frisiae*, the Frisians should also have been the descendants of the peoples, inhabiting the island situated not far from India where they had been very early converted to the Christian faith. From (over) there, the three brothers Friso, Saxo and Bruno departed with their families to the North. After a long wandering, they came to the Frisian shore where Friso – following the report – had given one part of the country (Du. *Zeeland*) to each of his seven sons under a separate rule. In accordance with this story, the Lord Almighty had special relations with Frisians, similarly as in earlier times with the people of Israel, so that the conversion of Frisians in the 8th century was in fact a renewed conversion.

The history of Frisians was modified, for instance, by the humanist Worp van Thabor who also claimed that the people in question came from Asia. Frisians from this account here descended from the army of Alexander the Great, marching from Asia Minor to the East. One part of this army should have conquered quite the whole of Asia, and after the death of its leader returned on three hundred vessels to Europe. The majority of them perished on the way home, and only fifty four arrived to the northern coastal regions, and twenty-one of them – to Saxony.

Frisians were depicted in epic texts often as allies of Christian princes in their fight with evil, and *Frisia* as a kingdom. Already in the epic texts from the 12th and 13th century, Frisian kings appeared, mostly as vassals of the Frank princes. Among them, as a well known personality, the legends mentioned Gondebald – the mythical king or Radbod (Redbad) – the historical king.

Furthermore, a story was handed down about how Charles the Great gave freedom to Frisians. It found among others its reflection in the work of Jacob van Maerland from the 13th century – *Spiegel Historial*. Frisians, under the command of Magnus, helped Charles the Great to capture Rome and therefore received from him different privileges (more about this event see Janse 1997: 77ff).

Known are also, for example, the legends, in which motives connected with Christendom appear, as, e.g., Legends of King Redbad (see Piebenga 1954, "Sêgen fan kening Redbad"). Once upon a time, Wilfryd, the archbishop of York from England, set out on a trip to Rome, and, during his voyage across the see, the storm had thrown his vessel upon the territories of Friesland, so that he found himself at the house of Aldgillis, the king of Frisians. Thus, Wilfryd was preaching the gospel in Friesland, and the Frisians accepted baptism from his hands. However, after the death of Aldgillis, when Redbad assumed power, missionaries had been driven out, the churches had been destroyed and paganism had been restored. The next story about Redbad told also that (once upon a time) he had to fight against Charles, the king of the Franks (Fri. *Karel fan Frankelân*). Before the battle started, negotiators agreed that both kings should stand on a piece of ground, each of them on one foot, with no arms. Who was able to stand longer should have received Friesland, and who put his foot on the ground had to leave Friesland forever. Promised to keep their laws and freedom, the Frisians were obliged to recognize the winner as a king. In the event that two kings stood on one foot the whole day and the whole night, Charles deliberately dropped his glove and Redbad lifted it up for his enemy, but it was impossible without putting a second foot on the ground. In consequence, Redbad had to leave Friesland. Afterward, Charles, as a ruler of Friesland, pointed out the place where the Frisians might gather at a folk meeting (the so-called *thing*) and ordered them to pass their own laws. The Frisians elected twelve "law-speakers" (Fri. *rjochtsprekkers* or *aesgen*) whose duty was to deal with the law. Since there was no results for a long time, Charles lost his temper and decided to punish them. The law-speakers were sentenced to death but they did not have to die immediately being sent to see on a ship with no sails or oar, no rudder or ropes. One of the judges asked then, why did they not pray for a thirteenth one who – like Jesus coming, even through the closed doors, to his disciples consoling and schooling them – could teach them their own law and would safely bring them to the shore. When the judges preached, they noticed a thirteenth man sitting on their ship and steering them with a great wooden log towards the shore. The man rescued them and taught them what the law is and what is not. A bit later, the number of *aesgen* returned to twelve, but from that time on there must have been a symbolic thirteenth judge in the country that taught the law at the place where the twelve judges had once reached the shore. Another story of that kind was about Redbad upon whom the miraculous salvation of a boy sentenced to death had made such a great impression that he considered whether or not to receive baptism. However, at the same time the devil started to tempt him. Redbad, standing on one foot in the baptismal font, concluded that those who had been not baptized and notably his ancestors – Frisians – could not be in the haven, so he decided to withdraw.

Myths as well as legends based on historical events had a symbolic value for Frisians, who were distinguished among the neighboring German tribes by their own law. They allowed the members of this ethnic group to identify with their own kinsmen and their culture. Among others, the myths on the origins of Frisians occupied a unique place. Moreover, motives connected with Christianity were often included in them. However, they symbolized, as such, the idea of Frisian freedom.

To be stressed also is that the history of Frisians has steadily inspired the bearers of Frisian as a means of communication to strive for equal rights. In various situations, they implicitly or explicitly refer to their own past. For example, starting from the 40s of the 20th century Frisians usually celebrate the battle at Warns (1345), in which the Frisians overcame the Dutch. Today, however, one can presume that the awareness of the distinct history among Frisians living in the province of Friesland is not so great. It results mainly from the fact that the history of the Netherlands as the history of the native country is taught in elementary schools.

2. National consciousness, "pillarization" and secularization of Frisian society

It is noticeable that the essence of Frisianness is not manifested in principle through adherence to political organizations or through religion. Frisians in the Dutch province of Friesland do not constitute a compact group as regards political and religious matters. The Frisian National Party (Fri. *Frysk Nasjonale Partij*), the representatives of which take part in the sessions of communal councils, and in the Councils of Provincial States, is not important, as to the number of its members and its spheres of influence (cf. Gorter 1997a: 1152ff). For example, the religious group of Mennonites (originally Anabaptists, named after Menno Simons, Frisian religious leader) being numerous in the 16th and 17th century constitutes only a minority within the Frisian population.

In this context, one can point at the phenomenon of "pillarization", i.e., segmentation of the society along religious and political lines. Similarly as in other countries, the citizens in the Netherlands are divided into the following four pillars: Protestant-Calvinistic (also with internal compartmentalization), Roman Catholic, Democratic Socialistic and Liberal. In the opinion of sociologists, the pillarization of the Dutch society – connected, since the 19th century, with the process of modernization in the fields of politics, economics and culture – was so significant, that it bestowed a second identity to individual citizens. The membership in one of these pillars seemed to be even more important than national identity. The succeeding process of secularization, which aimed at the depillarization, had taken place within the Dutch society until up the 70s and the 80s of the 20th century. The same societal changes had also reached the province of Friesland where the orthodox Protestant (Reformed Protestant) pillar was of great importance. In fact, 23% of Frisians belonged during the 20th century to the Reformed Protestant denomination. Towards the end of the 19th century, many Frisians being exploited by their Dutch employers decided to leave the Dutch Reformed Church that neglected the poor. In this sense, the withdrawal from the Dutch Reformed Church had something in common with the Frisian identity. It has been suggested by sociologists that Frisians are prone to cling to traditional principles and therefore prefer to accept solutions that are more radical. Hence, they rather decide to chose a radical breach with the Dutch Reformed Church and become churchless or, alternatively, to belong to the more strict Reformed Protestant denomination (cf. Jansma 1997b: 245ff).

3. Language as a fundamental symbol of Frisian ethnicity

The Frisian language has always been considered as the most important symbol of Frisian ethnicity. This fact may be illustrated by the tale about the Frisian hero – Grutte Pier. In 1515, Frisians found themselves under the rule of Saxons. Frisian villages and monasteries were plundered and destroyed, and people were driven out of their houses. Grutte Pier had also lost everything. Together with his cousin, Grutte Weird, he stood at the head of the army with 600 man which fought against the Hollanders, Burgundians and Saxons. At last, they defeated their enemies. As it had been reported, it was the Frisian language which constituted the main exponent of Frisianness for Grutte Pier. A story was handed over that who could not repeat the sentence: "Bûter, brea en griene tsiis, hwa dat net sizze kin, is nin rjuchte Fries" (Eng. 'Butter, bread and green cheese, who can't say it, is not a true Frisian'), or who pronounced it incorrect, had to die from his hands (cf. Dijkstra 1987 /1892–1896/: 65–67).

Nowadays, the language situation in the province – where, apart from Frisian and Dutch, numerous Frisian and non-Frisian (Dutch and Saxon) dialects appear – is rather complicated. From the viewpoint of sociology, Frisians in the Netherlands form today a language minority. Basing on the representative samples of investigative material collected by sociologists of language, one could state, how the Frisianness is connected with the Frisian language and how the attitude towards this language is manifested – intentionally or non-intentionally – through its use in interpersonal communication.

Not the whole population living in Friesland knows Frisian to the same extent, so that it is important in the context of Frisian identity to evaluate the degree of the knowledge of this language by the inhabitants of the province (cf. especially Gorter, Jonkman 1995: 7ff). The results of sociological questionnaires have shown, among others, that ca. 94,3% of the inhabitants of province understands Frisian, 74% can speak Frisian, 64,5% knows how to read in Frisian, and 17% knows to write in Frisian. Fortunately, the competence in the domain of Frisian language has not diminished starting from the 1960s in a drastic way, and generally, it has not changed from 1980. Moreover, it has been stated that the individuals with higher education, can relatively more often write in Frisian (27%), and people with lower education can more often speak Frisian (82%). However, linguistic abilities do not go hand in hand with the use of language in concrete communicative situations. Frisian vernaculars predominate in the informal domains, that is, in communication at home, with neighbors and friends. In the domains of public life, being more formal, Frisian functions only as a minority language.

The inhabitants of Friesland and the Frisian speakers possess certain images of their Frisianness. It has been noticed that the Frisian-speaking and Dutch-speaking individuals, as well as the autochthonous and non-autochthonous inhabitants of the province, while determining their ethnic identity, make use of different criteria (cf. van der Plank (1987: 17ff).

In the Netherlands, the Frisian identity is, in principle, treated as non-ethnic and in accordance with such assumptions every inhabitant of the province of Friesland is considered as a Frisian. In Friesland, however, genealogical or ethnolinguistic criteria are used for determining the ethnic identity of inhabitants. The bearers of Frisian usually declare their identity taking into account an ethnolinguistic criterion (that is, the mastering and the use of Frisian as a language of communication). The language is, in such a case, the most important factor that testifies to their affiliation with the ethnic group of Frisians. The Dutch-speaking autochthons in the province consider others and themselves as Frisians based on genealogical criteria.

In addition, it has been stated that the determination of ethnicity in accordance with linguistic criteria, connected with the emancipation of vernaculars, has found its support in Frisians from the lower strata of society (cf. in particular van der Plank 1987: 18–19). Thus, in the determination of identity such factors are taken into consideration as language and ethnic origin. Taking for granted that the spoken language plays a crucial role in their life, individuals speaking Frisian at home have identified themselves in 97% as Frisians. Those, however, who speak Dutch at home have considered themselves as Frisians to a lesser degree, i.e., in 41%.

Based on investigative surveys (cf. Gorter 1997b: 287–288), sociological polls have shown that 74% of all inhabitants of the province consider themselves as Frisians. Among them, those who use Frisian at home form 97%, those who are born in Friesland – 88%, those who live in the villages – 80%, those who have attained the age of 60 years – 81%, those who have a lower education 80%. The above statistics have shown how sociological and demographic data are decisive for drawing conclusions about the affiliation of individuals to the ethnolinguistic group of Frisians.

As to the question regarding their ethnicity, the inhabitants of Friesland have identified themselves in 36% as Frisians, 26% as Dutch, 11% as Frisian Dutch, 8% as Dutch Frisian, 3% as citizens of the world. 2% of the respondents have answered that they are Europeans, and 15% have claimed to have another identity. The results of these direct polls (cf. Gorter 1997b: 289f) have shown that the majority of people declare their adherence to the ethnic group of Frisians. Because there are those who view themselves as Frisian Dutch or Dutch Frisians, the boundaries between the Frisian identity and the Dutch identity is not so clearly cut.

One can assume, however, that the use of Frisian in concrete communicative situations is a means of expressing – intentionally or non-intentionally – Frisianness or solidarity with the Frisian minority. Therefore, one should quote exact numerical data. The studies of Gorter and Jonkman (cf. 1995: 24–26, and 72) have shown, for example, how high the percentage of the use of Frisian in comparison to other languages (by native speakers vs. non-native speakers) is. Proportionally to 100% of all its respective uses, Frisian is more often spoken in conversations: with a salesman in the store – 85% vs. 42%, close neighbors – 83% vs. 41%, and clerks at the regional post office – 80% vs. 32%. It occurs interchangeably with Dutch and non-Frisian dialects: in schools at the parents- and-teachers meeting – 76% vs. 31%, and in the contacts with a local policeman – 71% vs. 23%. Quite the same refers to the choice of a conversational language: by a foreigner at the door – 68% vs. 21%, an administrator of a group of villages in the communal offices – 60% vs. 22% or a doctor paying a visit to his patient – 56% vs. 18%. Rarely, the choice of Frisian is expected from a foreigner in the town – 41% vs. 13%, a doctor in the hospital – 22% vs. 5% and a Dutch-speaking neighbor living there for a year – 19% vs. 4%, a Dutch tourist who asks the way – 4% vs. 1%.

4. The ecological and societal-cultural symbols of Frisianness

Today, apart from the language, which separates and at the same time integrates the Frisians as an ethnic community, there are also other elements marking their distinct character, such as the Frisian culture or the landscape of Friesland and, especially, its localization in "Low Countries". Therefore, the term *it Frysk eigene* ('the Frisian intimate character') has been coined in the sociological literature on Frisian identity. It refers, in particular, to such core elements of Frisianness as the country folk with its peculiar character, habits and customs, history, culture, as well as the agrarian structure, the rural convivial atmosphere and the specificity of the landscape (cf. Gorter 1997b: 287). In the eyes of Frisianists and country patriots, Friesland is commemorated as a monument. Many villages expose the markers of their age. In numerous parochial churches, the reference to medieval times is made noticeable, and the names of the localities and streets are considered as testimonies of the rich past of this region (cf. Breuker, Janse 1997: 20). Moreover, the ethnologists (cf., e.g., van der Molen 1974: 3ff) mention within the concept of *it Frysk eigene* the Frisian habits and customs, the elements of material culture and art (the typical Frisian peasant-house, pieces of furniture, utensils, and the regional dresses of Frisians). The elevated clay mounds, on which Frisian established their settlements in the ancient period (the so-called *terpen*), occupy a remarkable place among typical signs of Frisian history. Furthermore, the houses with reinforced construction, built by Frisians since the Middle Ages to protect them against Norman invasions and floods (the so-called *stinzen* singular: *stins*), are also treated as a prominent exponent of the Frisian way of life.

Frisianness is identified – mainly in the Netherlands – with the peasant culture of the past. It is possible to find evidence (cf. Jensma 1998) that the creation of the picture of Friesland as a rural culture can be ascribed to its urban elite of the 19th century. Descending mainly from the

Frisian towns, intellectuals were afraid to loose the regional separateness of their culture as a result of social changes that took place in the Netherlands. Therefore they formed the so-called "folk culture", a kind a modern movement destined mainly for the townspeople. In accordance with the latter understanding of the Frisian culture, Frisianness happens to be expressed in the same manner today.

To illustrate the rural heritage of the urban culture in Friesland, it is sufficient to invoke the symbolic meaning of the title of the book *Het rode tasje van Salverda. Burgerlijk bewustzijn en Friese identiteit in de negentiende eeuw* written by Jensma (1998). In his studies on the Frisian identity in the 19th century, Jensma (1998: 13ff) purposefully alluded to Jan Cornelis Pieter Salverda and the literary episode making reference to his death and the unexplainable disappearance of his poems written in Frisian. The episode tells us about how Pastor Joost Hiddes Halbertsma – *nota bene* a poet with education – elevated the personality of a peasant poet, Salverda, to the symbol of Frisianness, giving a memorial speech in honor of him at the session of the Frisian Association (Du. *Fries Genootschap*). Salverda did not belong to the Frisian Association (having been created in 1827), the goal of which was to study Frisian culture and history. Its members were recruited mainly from higher ranks of Frisian society, the noble patrician class, statesmen, academic teachers, mayors, attorneys and judges. The poetry of Salverda had moved them all, but they considered his life stile as disgusting, because he stank unpleasantly. What Halbertsma exposed in the personality of Salverda was on the one hand his talent and on the other his poverty, deviated from accepted conventions. As it was reported (cf. Jensma 1998: 19), Salverda always carried with him a red briefcase, in which he kept his own poems. He used to repeat to his daughter Tiete: "When I die, you had to take this briefcase for you." After the poet gave up the ghost, the daughter looked for the red briefcase everywhere, between the books, in the coat, with which he was covered, in his last days on the bedding. Probably, it must have been thrown out with the dirt and numerous fleas from his bed and found itself on the heap of dangle.

Consequently, the events, which contributed to the formation of the position of Frisian as a means of communication in the 19th century, became the symbols of Frisianness. Starting from 1945 Frisians celebrate every year the anniversary of the battle at Warns (1345). This battle, for example, is seen as a symbol of resistance against Dutch authority. The performance of a biblical drama *Simon* (1948), written in Frisian by Fedde Schurer, in the capital of the province Leeuwarden (Fri. Ljouwert), belongs also to significant events of national character. As significant one should mention also the happenings on the streets of Leeuwarden in 1951, known as *Kneppelfreed* (Cudgel Friday), when Frisians demanded their rights were driven away with cudgels and fire extinguishers, are kept in precious historical memory. All such reported incidents symbolize an apotheosis of the fight for the priority of the Frisian culture in opposition to the Dutch culture (cf., e.g., Breuker, Janse 1997: 10). As tangible evidence of their national consciousness, Frisians have erected several monuments commemorating these and other important events or heroes.

The rise of Frisian consciousness has found its expression in the creation of organizations the aim of which is to cultivate the Frisian language and culture. The Frisian movement has existed in an organized form since the beginnings of the 19th century. Until the present time, the maintenance of Frisian as a language of education and public communication, being represented in literature and press, has strong institutional support from the authorities in the province of Friesland (cf. Gorter 1997a: 1152ff).

As for other exponents of Frisianness, competitions in sporting games belong to symbolic events on a national scale. For Frisians – who at least for two hundred years have striven to be the best ice skaters (cf. especially Breuker, Janse 1997: 59–61) – the so-called (in Fri.) *Alvestêdetocht*, i.e., the ice-skating race on the route uniting eleven Frisian towns is the most important game. At present, Friesland is represented also by such disciplines of sport (described in a special booklet: *Die Symbole von Friesland*), as: sailing on coastal ships with one mast (barge-sailing) (Fri. *skûtsjesilen*), pole-vaulting (Fri. *fierljeppen*), searching for eggs (Fri. *aeisykjen*), flicking up the ball (Fri. *keatsen*, cf. Fr. *jeu de pelote*).

The consciousness of ethnic separateness manifested by the inhabitants of the province of Friesland in the Netherlands finds its expression in the use of traditional symbols that customarily pertain to sovereign states, such as the flag, emblem and anthem. The flag and the emblem of Friesland have a common origin and refer to the early Middle Ages, the time when the seven independent Frisian countries situated along the coast regions of the North Sea joined in an alliance aiming at the common defense against Normans. The Seven leaves of the water lilies (Fri.

pompeblêdden) on the Frisian flag and seven quadrangles in the Frisian emblem constitute symbols of those seven Frisian countries. The Frisian flag contains the following elements: seven strips of the same breadth, placed aslant to the right upper corner, alternately cobalt-blue and white as well as seven dark red leaves of water lilies located on the white strives in the order of 2 – 3 – 2. The Frisian emblem includes two golden marching lions against an azure-blue background and seven golden rectangles in the order of 2 – 2 – 3. At the upper margin of the shield, a golden crone lies with five leaves and four pearls.

As regards the history of official symbols of Friesland, one should add that the Frisian flag with leaves of water lilies is known since the 11th century. In its present shape, it was accepted in 1897 by a special Dutch provincial commission, and in the year 1927 it was used officially for the first time. A look into heraldic books from the 15th century reads that an originally single emblem with water lilies and a lion has served as a basis for creating two new emblems. The first emblem exhibits two lions and seven small rectangles that have been transformed from the leaves of water lilies, and the second one includes (solely) strives and leaves of water lilies. The latter, in its final shape, was used as an official stamp of the Frisian Association (Du. *Fries Genootschap*) (since 1830). Moreover, this specimen of a national symbol was in turn accepted at the end of the 19th century as a Frisian flag.

Other kind of ethnic exponents that have traditionally integrated the members of this speech community constitute Frisian folksongs. Among them, the song *De âlde Friezen* ('Old-Frisians'), sung on the melody of the German student-song "Vom hoh'n Olymp", which praises Friesland as "the best land in the world" (Fri. *it bêste lân fan d'ierde*), is the most significant. This song had been accepted as a kind of national anthem by those who assembled on the occasion of the erection of the monument of Eeltsje Hiddes Halbertsma, the poet and the author of its words, when it was performed in the year 1875. From that time on, the Old-Frisians' song became widely known as "Frysk Folksliet". It was replaced in this particular role, mainly after the World War II, by a song from the beginning of the 20th century, entitled *It Heitelân* ('The Fatherland') with the music written by J. Lindeman. It is usually sung with the same willingness by Frisians as: *It Heitelân*, or *De Wâldsang* ('The song of the forest'). (A detailed description of the Frisian flag, emblem and anthem is to be found in the booklet *Die Symbole von Friesland*).

5. Summarizing the semiotic markers of Frisian ethnicity

To sum up, Frisians as a primordial people have preserved their land, traditions, customs, and modes of existence until today. If we try to approach the signs of their ethnic identity from the viewpoint of semiotics, we can state that various elements have the meaning of symptoms (indicators), appeal signals and symbols of ethnicity. We may emphasize that the ethnic character of Frisian society becomes transparent if we order its societal facts in accordance with their importance for the group identification.

Thus, the Frisians who form at present a consolidated speech community using their own native language in every day communication consider it as the most important symbol of Frisian ethnicity. In the Dutch province of Friesland, a symbolic function, although to a minimal extent, is fulfilled by the standard variety of Frisian. Furthermore, the use of the Frisian language is perceived as an expression of belonging to a given speech community or plays a role of a signal evoking ethnic solidarity or language loyalty of the interlocutors in concrete communicative situations.

The national values are not only embodied in the native language of Frisians. In the past, Frisian mythology and the distinctness of Frisian law arose to a symbolic dimension. At present, the separate character of Friesland is expressed through a distinct Frisian culture or landscape, formed by the activity of people inhabiting the Frisian territory. As marks of ethnic identity, one considers, e.g., cloths, customs and habits. As in the past, the Frisianness is marked by a specific Frisian mentality. It always expresses itself in the acceptance of extreme social attitudes, as, e.g., recently, in the choice of not belonging to any church or belonging to a strict Reformed Protestant denomination.

Among the different markers signaling the affiliation to distinct groups of Frisians, one can enumerate different forms of both non-verbal and verbal behavior, including those of professional activity in favor of Frisian-speaking communities, as well as cultural artifacts, handcraft and literary works, and the like. Traditional signs of nation and history (e.g., myths and legends) slowly have lost impact and importance; being replaced by a growing number of new expressions of regional and ethnic identity, as, for example, sports (in particular *Alvestêdetocht* 'Eleven Cities Tour').

Supplement Three

Presenting West Frisians as an aggregation of ecological linkages

Human linguistics as an investigative perspective has been introduced to Poland at the workshop, "Exploring the Domain of Human-Centered Linguistics from a Hard-Science Perspective", organized by Victor Huse Yngve and Zdzisław Wąsik (2000). As a consequence of subsequent working conventions at the Societas Europea Annual Meetings and the Linguistic Society of the United States and Canada Forums within the next four years, its results were published in the form of an academic handbook *Hard-Science Linguistics* (Yngve and Wąsik 2004).

Provided with an introductory and the concluding frame of reference this paper presents a report on philological investigations conducted in the School of English at the Adam Mickiewicz University in Poznań in accordance with the principles of V. H. Yngve's book *From Grammar to Science* (1996). This report recapitulates the achievements of the author's studies (Wąsik, E. 2000, 2004) analyzing the formation of linguistic communities of Frisians in terms of task-oriented long-lasting linkages.

1. Applying Victor H. Yngve's theory of human linguistics to the analysis of interpersonal relationships among West Frisians

The objective of E. Wąsik's studies (2000, 2004); popularized lately in a joint publication of Z. Wąsik, E. Wąsik and Maciej Kielar (2006) was a typological survey of the linguistic-communicational means that signify the national identity of West Frisians living in the Dutch province of Friesland. Fundamental data for her studies, conducted in accordance with the principles of human linguistics had been excerpted from sociological and historical investigations (among others Gorter et al. 1984; Breuker and Janse 1997; Jonkman 1997; Piebenga 1954). The distinctiveness of Frisians, however, except for the use of their language constituting a core exponent of ethnic identity, has been revealed also by other observable parts of linkages, considered in human linguistics as channels, props or settings.

It has been stated that West Frisians today form a relatively compact communicative minority within the state of the Netherlands, being related also to North Frisians and East Frisians living in the Federal Republic of Germany. They have preserved their separateness from the Dutch-speaking inhabitants in the province of Friesland, and generally from the remaining inhabitants there, to the extent that their language, standardized Frisian, is spoken today as the second official language in the Netherlands. In the light of human linguistics, the community of West Frisians can be seen as a set of coupled linkages with various central and peripheral linguistic properties forming through the ages a large and long-lasting linkage associated with a geographical, political and/or cultural surrounding. The number of the participants in the West Frisian linkage is at present about 400 000 if one considers those inhabitants of the province who can speak Frisian. However, the total population of the province, as it was mentioned earlier, amounts to 600 000 (according to the statistical data of Gorter 1997a: 1152).

Frisians were traditionally recognized as a distinct ethnic group, at least within the province, because they knew Frisian and spoke it in various communicative situations. Knowledge of Frisian against the background of Dutch was considered as the most important property of their identity. However, in recent times language does not appear as an important category for the affirmation of Frisianness as, for example, place of birth (more about the criteria of self-determination see in Gorter 1997b: 287–288).

How Frisians managed to preserve their ethnic identity over the ages might be answered against the background of the concept of tasks (and their subtasks) explaining the communicational activity of individuals in social linkages (cf. Yngve 1996: 186, 265). In totality, the main task of the West Frisian linkage as a whole has always been to keep their separateness as a group. At present, as in the case of many ethnic or linguistic minorities, the principal task of West Frisians is to strive for the equality of their rights in relation to the majority group within the state. They have striven to possess a unique position within the framework of Dutch culture, referring in this respect to their own history and their own verbal forms of expressing communicational tasks (cf., e.g., Breuker and Janse

1997: 11; Jonkman 1997: 258). The subtasks of groups and particular linkages communicating in Frisian have in great measure been subordinated to the main task of expressing their ethnic identity.

The transmission of an ethnic identity of Frisians through the interaction of individuals and linkages was due to the fact that in the early Middle Ages, they were an important sea-faring and trading people. To the common tasks of Frisians belonged their struggle with the sea. At that time the formation of long-lasting types of linguistic linkages focusing on the realization of social tasks was connected with the person-to-person communication. In particular, it was the Old Frisian law which was transmitted through oral tradition for ages. However, Frisians were distinguished from neighboring German tribes by written forms of law.

Speaking in terms of human linguistics, as props that linked Frisian communities from the ancient time, one has to consider runic inscriptions (from the 6th–8th century) and above all the oldest of the preserved Old Frisian sources, i.e., legal texts (from the 13th and 14th centuries). Starting from the 17th century, literary works written in Frisian, first in the West and then in other parts of Frisia, have played a linking role as props.

Among personalities uniting many Frisians into coupled linkages one has to consider the religious leader Menno Simons. However, Mennonites (originally Anabaptists) being numerous in the 16th and 17th century constitutes today only a minority within the Frisian population.

Myths of the origin of Frisians and stories about historical events referring to Frisians and Frisian heroes can be treated as an example of "rumor-kind" messages uniting the Frisian linkage in time and space. As "focused" linkages, one can regard small groups with fuzzy boundaries to which such legends were transmitted orally. The groups of persons linked by stories could be extended to the scope of complete linkages. Important also are the settings in which these legends and myths had been created, their creators, authors of works from the period of the Middle Ages and the Renaissance, who amended and modified these stories so that they won broader and broader circles of receivers. One could also include here mythical or historical personalities, as, e.g., the king of the Frisians, Radbod (Fri. *Redbad*), the king of Franks, Charles the Great, and the national hero, Magnus. Similarly, among props or channels one could enumerate the written works, within which these stories were published, and also paintings, pictures or other illustrations of personalities or events mediating between their creators and receivers.

To the legends distributed among Frisians in a "rumor-kind" of linkages belongs an old story relating how travelers who arrived by sea from Asia to the northern countries became founders of new peoples. Famous among Frisian personalities was a certain Friso, who had seven sons, and who gave each of them one part of Frisia to rule. Furthermore, a story was handed down about how Charles the Great had given freedom to the Frisians. It found reflection in the work of Jacob van Jacob van Maerlant of the 13th century (*Spieghel historiael*, began 1283), a rhymed chronicle of the world, translated from the *Speculum historiale* of Vincent of Beauvais. Frisians under the command of Magnus had helped Charles the Great to capture Rome and therefore had received from him diverse privileges (see Janse 1997: 77ff). To the Frisian tradition belong also the legends of the "rumor" kind, i.e., repeated from mouth to mouth, in which motifs connected with Christendom appear, as e.g., Legends of King Redbad (see Piebenga 1954). They tell the stories of: (1) how Frisians accepted baptism from the hands of Wilfryd, the archbishop of York from England, (2) how Redbad while having assumed power expelled missionaries from the country and ordered the destruction of the churches and the restoration of paganism, (3) how Redbad had to fight against Charles, the king of Franks, (4) how Charles, as a ruler of Friesland, pointed out the place where the Frisians might gather at a folk meeting (the so-called *thing*) and ordered them to pass their own laws, and (5) how Redbad, standing on one foot in the baptismal font, decided to withdraw from the rite of baptism when he realized that those who had not been christened before, and particularly his ancestors, could not enter the kingdom of heaven.

In the historical context of political and religious life in Friesland, one can point to the phenomenon of "pillarization", i.e., the segmentation of society, which has assumed importance there since the 19th century. As a result of modernization in the domain of politics, economics and culture, the citizens were divided, as in the Netherlands, into four pillars: Protestant-Calvinistic (also with internal compartmentalization, as Orthodox Reformed Protestant and Liberal Protestant), Roman Catholic, Democratic Socialistic, and Conservative. In the province of Friesland, the Orthodox Reformed Protestant pillar was of great importance. The pillarization of society means that a person being born as a Protestant-Calvinist was sent to denominational schools from kindergarten to university. Furthermore, he became a member of a denominational trade union and political party; he read newspapers, listened to radio programs and watched television programs subordinated to the same ideology. Consequently, he would spend his free time (sports and recreation) as a member of a denominational association (cf., e.g., Jansma 1997: 245ff).

The pillarization of Frisian society, as a typical phenomenon in fact, had a great influence on the formation of linkages on various levels of social life (including examples of interaction through arrangement). Important, in the same context, was the succeeding process of secularization which took place within Dutch society up to the 70s and the 80s of the 20th century and and which ended with depillarization.

Relating to the present situation, Frisians can be seen as functioning within various subordinate, "coupled" and mutually conditioned linkages. The bases for the distinction of typical Frisian linkages create such traditional settings of language communication as work, school, and public life. However, the essence of Frisianness does not manifest itself only through the adherence to political organizations or through religion. Actually, political organizations or political parties function in the political and religious life of Frisians as large and long-lasting linkages where the Dutch-speaking and Frisian-speaking participants of communication meet together. There is a noticeable relationship between linguistic background and political preference. Small conservative religious parties, like the Frisian National Party and the Christian Democratic Party, are supported mostly by Frisian-speaking members. In reality, these parties constitute also the most autochthonous parties. The Social Democratic Party has an equal number of Frisian and non-Frisian speakers. The Frisian National Party, although not important as to the number of its members and its spheres of influence in the light of Gorter's study (1997a: 1152ff), can be considered as a specific Frisian linkage. Its main task is to support the existence of Frisian in local communication, e.g., in the sessions of communal councils, and in the Councils of Provincial States.

Among Frisian-speaking groups, it is hard to distinguish large and long-lasting academic linkages, although Frisian studies are represented at some larger universities in the Netherlands. Members of the *Fryske Akademy*, existing since 1938, are linked through the tasks of organizing scientific conferences devoted to the Frisian language, literature, and culture. One can encounter also professional organizations, schools, corporations, or trade unions where the employees meet regularly to perform their common tasks of promoting the idea of Frisianness. In the province of

Friesland, there are two types of bilingual linkages that are coupled through education, namely, Frisian-speaking and Dutch-speaking linkages. In recent years, the majority of young inhabitants of the province has an opportunity to be educated in Frisian at the primary and secondary school level.

In the cultural life of the province, common enterprises on a national scale contribute to the formation of social groups of temporary and long-lasting linkages. They focus on the organization of such social gatherings as sporting events, musical or theatrical performances, etc. Frisians participate in such typical group sports as sailing on coastal ships with one mast, pole-vaulting, searching for eggs, flicking up a ball. Brass bands and choirs and several Frisian hard-rock groups and amateur theatre groups play and sing in the Frisian language. Organizations of all sorts – including writers, correspondents, readership associations of newspapers or magazines, and performers and audience of a particular theater or stadium, are flourishing in villages, towns, cities, and even their performances are currently available on radio or television.

Following sociological studies (cf. Gorter and Jonkman 1995: 24–26, and 72), several situations have been distinguished where everyday understanding in the Frisian language takes place. They contribute to the formation of small and brief or longer-lasting linkages among Frisians and Non-Frisians, e.g., in conversations with a salesman in a store, between close neighbors, with clerks at a local post office, between parents and teachers at a school meeting, with a district policeman, with a foreigner at the doorstep, with an administrator of a group of villages in the communal offices, with a doctor paying a visit to his patient, with a foreigner in the town, with a doctor in the hospital, or with a Dutch-speaking neighbor living there for a year, as well as with a Dutch tourist who asks the way.

Sociological data provide an image of the extent to which the inhabitants of the province use Frisian in a number of settings along with social properties of the communicating individuals connected with the choice of Frisian. Generally, the use of Frisian by the inhabitants of the province decreases in the more formal domains of public life. It depends on: (1) whether Frisian has been learned as a first language at home or as a second language, (2) whether the attitude towards Frisian is positive or not, and (3) whether the speakers of Frisian live in the country or in a town. In particular, born-and-bred Frisian speakers having a higher occupational position or who are younger use Frisian relatively less. The difference in the use of Frisian is greatest between the rural population and the inhabitants of the cities partially as a result of the relatively smaller number of Frisian speakers who live in cities. Frisian speakers themselves use their language less in an urban environment, particularly in urban public life and in mixed neighborhoods. The lowest use of Frisian is in the provincial capital Leeuwarden. Frisian-speaking youths are strongly underrepresented at higher levels of higher education, but more pupils speak Frisian or another dialect with each other than have acquired these as first languages at home. Frisian speakers use their language at work relatively rarely. Many Frisian speakers at a higher level of social ranking do not speak Frisian with their colleagues, but above all, they do not speak Frisian with those above or below them in the hierarchy. At lower levels, a large majority of Frisian speakers do use Frisian with their co-workers. Frisian is used most often with customers. Frisian speakers and non-Frisian speakers are members of Frisian-speaking and non-Frisian-speaking associations respectively. The division is so sharp that one could speak of a linguistic border in associations and club life, determined additionally by the place of dwelling. Frisian speakers in a city are on the average less often members of a monolingual association, while Frisian speakers in the country are more often members of such associations. Remarkable diversity in language use is exhibited in public life. A great deal more Frisian is spoken by those employed in the assistance and service sectors that have a somewhat lower status than by persons who perform higher social functions.

Moreover, one can observe that more than half of the school-age children use Frisian with their playmates, but many more in the youngest age group use a combination of Frisian and another language than in the oldest age group. Along with education, public administration, and the media, religion can also be regarded as one of the highly developed domains of language communication. Although services in Frisian are only held sporadically, most of the inhabitants of the province who go to church have attended a Frisian service at one time or another. In many congregations, services in Frisian are held a couple of times a year. Of those who speak Frisian at home not even half appear to pray in this language. Only one fourth of those who are religious-minded appear to have a copy of the Frisian Bible translation (completed in 1943), only one fifth use it. It seems however, that the barriers between the language of home and colloquial language and the language of religion are slowly being broken down among young people, who are beginning to use Frisian in their religious life.

The media, radio, television and press, which can be regarded as channels in the communication between the smaller Frisian linkages, uniting them into a large superordinate linkage, play to a certain degree a unifying role in relation to participants of Frisian-speaking linkages. Although articles in Frisian newspapers occupy only five percent of the editorial space, they are read regularly by half of the readers. One third of the population of Friesland regularly listens to special radio and television programs in Frisian broadcast by the Regional Broadcasting Organization of Friesland. Frisian books and periodicals play the role of props in communication within the groups of long-lasting and brief linkages, when estimated from their supply in the market and the statistics of readership in libraries. Over 100 new books are published every year. Some books are translations (e.g. children books). Poetry and novels are mostly original Frisian books. Publishing in Frisian is stimulated by the provincial government and professional councils and is available through the Frisian Literary Museum and the Foundation for the Frisian book. Moreover, volunteers carry out an annual promotional campaign for Frisian books by means of door-to-door sales. The books are carried in wheelbarrows (also to be considered as props) through neighborhoods and villages. However, one can point to an important property of the participants of the Frisian linkage, namely that only 30% of the total population of Friesland can read in Frisian. As a consequence of this state of affairs, one can observe that as the purchases of Frisian books increased; the borrowing of Frisian books from lending libraries has decreased. Periodicals, which appear primarily or entirely in Frisian and whose primary circulation area is Friesland, have generally a Frisian cultural character. They are aimed at a small group. Sixteen percent of the Frisian population has occasionally read a Frisian-language periodical; the number of their subscribers is most certainly well below that.

Among the exponents of the ethnic identity of Frisians, one can also enumerate such properties of communicating individuals as, e.g., speaking in a standard variety or in a dialect, code-switching, linguistic interference, etc. Here, Frisian vernaculars constitute a core element marking the ethnicity of Frisians in addition to their folk habits and customs, history and culture defined next to such components as the agrarian structure, the convivial rural atmosphere and the landscape. To other factors, which embody the concept of Frisianness, belong also the nationality name, the myth of common descent, a unique culture, the association with a specific territory, the sense of solidarity, and intimacy.

To sum up, the application of the perspective of human linguistics enables researchers to present their knowledge about the society with respect to the changeable nature of multilingual linkages, which are conditioned by dynamic interactions among their members. Therefore, sociologist's data always have to be modified as regards their typological assessments.

2. Observable and non-observable facts in human linguistics: between hard and soft sciences

To sum up, one has to state that Yngve's theory of human linguistics can be applied in the search for the linguistic-communicational properties of man but only on the level of facts, which are observable. Verbal means, as, for example, objects that are linguistically relevant, and, in the case of articulated sound waves, being sent and received, have always to underlie the interpretation of receivers on the basis of their solipsistic experience and knowledge. In the physical domain, only the verbal means and their accompanying nonverbal contexts constitute objectively available components of reality giving one or another person the possibility of accessing the subjectivity of another person. Despite the assumed hard-science approach to human communication, researchers are not in a position to remain exclusively in the physical domain, as Victor H. Yngve claimed, in order to make inquiries into the linguistic properties of people from the viewpoint of physics, chemistry and biology. It is especially apparent in the search for tasks, which the participants of group communication fulfill. What kind of tasks can be ascribed to communicating individuals in particular or collectivities in general depends, as a matter of fact, on the subjective inferences made both by communication participants and the researchers who act in any case as receivers. The same thing may be said with reference to the addressees in communicational acts who interpret the referential value of linguistic utterances on the basis of introspection. It is obvious that tasks, similarly as subjective meanings or values, are only ascribed to phonic sequences or to other physical objects, when they are relevant from a linguistic point of view.

Bibliography

Works cited and consulted

Adamska-Sałaciak, Arleta 1996: *Language Change in the Works of Kruszewski, Baudouin de Courtenay and Rozwadowski*. Poznań: Motivex.
Ammon, Ulrich 1987: Funktionale Typen/Statustypen von Sprachsystemen. In: Ulrich Ammon, Norbert Dittmar, Klaus J. Mattheier (eds.) 1987, 230–263.
Ammon, Ulrich, Norbert Dittmar, Klaus J. Mattheier (eds.) 1987: *Soziolinguistik. Ein internationales Handbuch zur Wissenschaft von Sprache und Gesellschaft*. 1. Halbband. Berlin, New York: Walter de Gruyter.
Angell, James R(owland) 1904: *An Introductory Study of the Structures and Functions of Human Consciousness*. New York: H. Holt.
Angell, James R(owland) 1907: *The Province of Functional Psychology*. Baltimore, MD: Review Publishing.
Antas, Jolanta 2000 [1999]: *O kłamstwie i kłamaniu. Studium semantyczno-pragmatyczne* [On the lie and lying. A semantic-pragmatic-study]. 2nd edition. Kraków: Towarzystwo Autorów i Wydawców Prac Naukowych Universitas.
Austin, John L(angshaw) 1975 [1962]: *How to Do Things with Words*. The William James Lectures Delivered at Harvard University in 1955. 2nd edition. Edited by J(ames) O(pie) Urmson, Marina Sbisà. Oxford, UK, New York, NY: Clarendon Press.
Awdiejew, Aleksy, Krzysztof Korżyk 2000: Communicative grammar: Toward a linguistic model of interpretative activity. *Biuletyn Polskiego Towarzystwa Językoznawczego /Bulletin de la Société Polonaise de Linguistique* LVI, 15–37.
Bańczerowski, Jerzy 2001: The linguistic legacy of Ludwik Zabrocki. In: Stanisław Puppel (ed.) 2001: *The Ludwik Zabrocki Memorial Lecture*. Poznań: Wydział Neofilologii Uniwersytetu im. Adama Mickiewicza w Poznaniu, 9–49.
Bańka, Józef 1986: *Ontologia bytu aktualnego. Próba zbudowania ontologii opartej na założeniach recentywizmu*.[Ontology of actual being. An attempt at building an ontology based on the principles of recentivism]. Katowice: Uniwersytet Śląski.
Barker, George C(arpenter) 1947: Social functions of language in a Mexican-American community. *Acta Americana* 5, 185–202.
Barker, George C(arpenter) 1972: *Social Functions of Language in a Mexican-American Community*. Tucson: University of Arizona Press.
Barker, Roger G(arlock) 1968: *Ecological Psychology. Concepts and Methods for Studying the Environment of Human Behavior*. Stanford CA: Stanford University Press.
Barker, Roger G(arlock) et al. 1978: *Habitats, Environments, and Human Behaviour. Studies in Ecological Psychology and Eco-Behavioral Science from the Midwest Psychological Field Station*, 1947–1972. San Francisco: Jossey-Bass Publishers.
Bateson, Gregory 1972: *Steps to an Ecology of Mind*. New York: Ballantines.

Berger, Peter L(udwig), Thomas Luckmann 1966: *The Social Construction of Reality.* Garden City, N.Y.: Doubleday.
Berndt, Heide 1968: Ist der Funktionalismus eine funktionale Architektur? Soziologische Betrachtung einer architektonischen Kategorie. In: Heide Berndt, Alfred Lorenzer, Klaus Horn (eds.) 1968, 9–50.
Berndt, Heide, Alfred Lorenzer, Klaus Horn (eds.) 1968: *Architektur als Ideologie.* Frankfurt am Main: Suhrkamp.
Boissevain, Jeremy 1974: *Friends of Friends: Networks, Manipulators and Coalitions.* Oxford: Basil Blackwell, New York: St. Martin's Press.
Brentano, Franz 1874: *Psychologie vom Empririschen Standpunkt.* Leipzig: Teubner.
Breuker, Philippus H., Antheun Janse 1997: Beelden [Pictures]. In: Philippus H. Breuker, Antheun Janse (eds.) 1997, 9–66.
Breuker, Philippus H., Antheun Janse (eds.) 1997: *Negen eeuwen Friesland-Holland. Geschiedenis van een haat-liefdeverhouding* [Nine centuries Friesland-Holland: A history of a hate-love relationship]. Zutphen: Fryske Akademy & Walburg Pers.
Bühler, Karl 1965 [1934]: *Sprachtheorie. Die Darstellungsfunktion der Sprache.* 2nd edition. Stuttgart [Jena]: Gustav Fischer.
Bühler, Karl 1990 [1934]: *Theory of Language: The Representational Function of Language.* Translated by Donald Fraser Goodwin. Amsterdam, Philadelphia: John Benjamins [*Sprachtheorie. Die Darstellungsfunktion der Sprache.* Jena: Gustav Fischer].
Carr, Hervey A. 1925: *Psychology: A Study of Mental Activity.* New York, NY et al.: Longmans. Green & Co.
Carston, Robyn 2002: *Thoughts and Utterances. The Pragmatics of Explicit Communication.* Oxford: Blackwell Publishing.
Cassirer, Ernst 1944: *An Essay on Man: An introduction to the Philosophy of Human Culture.* New Haven: Yale University Press; London, H. Milford, Oxford University Press.
Chomsky, Noam (Avram) 1965: *Aspects of the Theory of Syntax.* Cambridge, MA: The Massachusets Institute of Technology Press.
Chomsky, Noam (Avram) 1986: *Knowledge of Language: Its Nature, Origin and Use.* New York: Praeger.
Christy, Thomas Craig 1983: *Uniformitarianism in Linguistics.* Amsterdam: John Benjamins (Amsterdam Studies in the Theory and History of Linguistic Sciences. Series 3: Studies in the History of Linguistics 31).
Collinge, N(eville) E(dgar) (Oscar) 1995a: History of comparative linguistics. In: E(rnst) F(riderik) K(onrad) Koerner, R. E. Asher 1995, 195–202.
Collinge, N(eville) E(dgar) (Oscar) 1995b: History of historical linguistics. In: E(rnst) F(riderik) K(onrad) Koerner, R. E. Asher (eds.) 1995, 203–212.
Danesi, Marcel 2002: *Understanding Media Semiotics.* London: Edward Arnold and New York: Oxford University Press.
Dawkins, Richard 1982: *The Extended Phenotype.* Oxford: Oxford University Press.
Deely, John (N.) 2000: Semiotics as a postmodern recovery of the cultural unconscious. *Sign Systems Studies* 28, 15–48.
Devetak, Silvo 1996: Ethnicity. In: Hans Goebl, Peter H. Nelde, Zdeněk Starý, Wolfgang Wölck (eds.) 1996, 203–209.

DeVito, Joseph A. 1976: *The Interpersonal Communication Book*. New York, Hagerstown, San Francisco, London: Harper & Row.
Dewey, John 1896: The reflex arc concept in psychology. In: *Psychological Review* III (July), 357–370.
Di Cristo, Albert 2000: Une grammaire écologique comme cadre interpretatif de la prosodie de la parole. In: *Abstracts. Nordic-Baltic Summer Institute for Semiotic and Structural Studies, 19th Annual Meeting of the Semiotic Society of Finland, Finish Summer School, June 12–21, 2000 in Imatra, Finland*. Imatra: Imatran kaupunki Monistamo, 17–21.
Die Symbole von Friesland. Leeuwarden: Ausgabe der Provinzialverwaltung von Friesland. Presse- und Informationsstelle 'Bureau Voorlichting' (a popular booklet without any publication date).
Dijkstra, Waling Gerrits 1987 /1892–1896/: Grote Pier [Grand Peter]. (Fragment from: Waling Gerrits Dijkstra 1892–1896: /Uit Friesland's volksleven van vroeger en later. Volksover-leveringen, volksgebruiken, volksvertellingen, volksbegrippen. Leeuwarden: H. Suringar/) In: Pieter Terpstra (ed.) 1987: Fries Letterland. Poëzie en proza op Friese bodem [Frisian Literary Land. Poetry and Prose on the Frisian Ground]. Amsterdam: Sijthoff, 65–67.
Dik, Simon C. 1981 [1978]: *Functional Grammar*. 3rd printing. Dordrecht: Foris [Amsterdam: North Holland].
Dik, Simon C. 1987 [1982]: Some principles of Functional Grammar. In: René Dirven, Vilém Fried (eds.) 1987: *Functionalism in Linguistics*, Amsterdam, Philadelphia: John Benjamins, 81–100 [In: Preprints of the plenary session on syntax, 13th International Congress of Linguistics, Tokyo, 66–76].
Downes, William 1998 [1984]: *Language and Society*. 2nd edition. Cambridge, UK, New York, NY, Melbourne, Australia: Cambridge University Press [London: Fontana Paperbacks].
Duranti, Allessandro 2000 [1997]: *Linguistic Anthropology*. Cambridge: Cambridge University Press (Cambridge Textbooks in Linguistics).
Edelman, Gerald M. 1992: *Bright Air, Brilliant Fire: On the Matter of the Mind*. New York: Basic Books.
Emmeche, Claus 2001: Bioinvasion, globalization, and the contingency of cultural and biological diversity: Some ecosemiotic observations. *Sign Systems Studies* 29 (1), 237–262.
Enninger, Werner, Karl-Heinz Wandt 1984: Language ecology revisited: From language ecology to sign ecology. In: Werner Enninger Lilith M. Haynes (eds.) 1984: *Studies in Language Ecology*. Wiesbaden: Franz Steiner, 29–51.
Erickson, Frederick 1987: Ethnicity. In: Ulrich Ammon, Norbert Dittmar, Klaus J. Mattheier (eds.) 1987, 91–95.
Fasold, Ralph W. 1984: *The Sociolinguistics of Society*. Oxford: Basil Blackwell.
Fasold, Ralph W. 1989: Naturalism and the search for a theory of language types and functions. In: Ulrich Ammon (ed.) 1989: *Status and Function of Languages and Language Varieties*. Berlin, New York: Walter de Gruyter, 107–121.
Fawcett, Robin P. 1980: *Cognitive Linguistics and Social Interaction. Towards an Integrated Model of a Systemic Functional Grammar and the Other Components of a Communicating Mind*. Heidelberg: Julius Groos Verlag.

Ferguson, Charles A(lbert) 1959: Diglossia. *Word* 15, 325–340.
Ferguson, Charles A(lbert) 1962a: The language factor in national development. *Anthropological Linguistics* 4 (1), 23–27.
Ferguson, Charles A(lbert) 1962b: Background to second language problems. In: Frank A. Rice (ed.) 1962, 1–14.
Ferguson, Charles A(lbert) 1966a: National sociolinguistic profile formulas. Discussion. In: William Bright (ed.) 1996: *Sociolinguistics*. The Hague, Paris: Walter de Gruyter, 309–324.
Ferguson, Charles A(lbert) 1966b: Sociolinguistically oriented language surveys. *Linguistic Reporter* 8 (4), 1–3.
Fill, Alwin (ed.) 1996: *Sprachökologie und Ökolinguistik. Referate des Symposiums Sprachökologie und Ökolinguistik an der Universität Klagenfurt, 27.–28. Oktober 1995*. Redaktionelle Mitarbeit Hermine Penz. Tübingen: Staufenburg.
Fill, Alwin 1993. *Ökolinguistik. Eine Einführung*. Tübingen: Gunter Narr.
Fill, Alwin, Peter Mühlhäusler (eds.) 2000: *The Ecolinguistics Reader: A Selection of Articles on Language, Ecology, and Environment*. London, New York: Continuum.
Firth, John R(upert) 1957 [1935]: The technique of semantics. In: J. R. Firth 1957. *Papers in Linguistics 1934–1951*. London: Oxford University Press, 7–33 [Transactions of the Philological Society 36–72].
Fishman, Joshua A(aron) (ed.) 1970 [1968]: *Readings in the Sociology of Language*. 2nd printing. The Hague, Paris: Mouton.
Fishman, Joshua A(aron) 1964: Language maintenance and language shift as a field of inquiry. *Linguistics* 9, 32–70.
Fishman, Joshua A(aron) 1965: Who speaks what language to whom and when. *Linguistique* 2, 67–88.
Fishman, Joshua A(aron) 1986 [1972]: Domains and the relationship between micro- and macrosociolinguistic. In: John J. Gumperz, Dell Hymes (eds.) 1986: *Directions in Sociolinguistics. The Ethnography of Communication*. Oxford, New York: Basil Blackwell [New York: Holt, Rinehart and Winston].
Foley, William A. 1997: *Anthropological Linguistics*. Oxford: Basil Blackwell.
Furdal, Antoni 1990 [1977]: *Językoznawstwo otwarte* [Open linguistics]. 2nd enlarged edition. Wrocław [Opole]: Zakład Narodowy imienia Ossolińskich.
Gardiner, Alan H. 1932: *The Theory of Speech and Language*. Oxford: At the Clarendon Press.
Garvin, Paul L(ucian), Madeleine Mathiot 1970 [1968]: The urbanization of the Guarani language: A problem in language and culture. In: Joshua Fishman A. (ed.) 1970, 365–374.
Gawroński, Andrzej 1928 [1927]: O istocie i rozwoju języka [On the nature and development of language]. In: Andrzej Gawroński 1928: *Szkice językoznawcze* [Linguistic sketches]. Warszawa, Kraków, Lublin, Łódź, Poznań, Wilno, Zakopane: Gebethner & Wolff, 1–37 [La langue, sa nature et son origin. *Biuletyn Polskiego Towarzystwa Językoznawczego* 1, 3–93].
Giddens, Anthony 1990: *The Consequences of Modernity*. Stanford, CA: Stanford University Press.
Givón, Talmy 1984: *Syntax, a Functional-Typological Grammar*. Amsterdam: John Benjamins.

Givón, Talmy 1995: *Functionalism and Grammar*. Amsterdam, Philadelphia: John Benjamins.
Glasersfeld, Ernst von 1988: The reluctance to change a way of thinking. *The Irish Journal of Psychology* 9 (1), 83– 90.
Glasersfeld, Ernst von 1995: *Radical Constructivism: A Way of Knowing and Learning*. London, Washington: The Falmer Press.
Glasersfeld, Ernst von 2001: The radical constructivist view of science. *Foundations of Science, special issue on "The Impact of Radical Constructivism on Science"*, edited by Alexander Riegler, vol. 6, no. 1–3, 31–43.
Goebl, Hans, Peter H. Nelde, Zdeněk Starý, Wolfgang Wölck (eds.) 1997: *Kontaktlinguistik/Contact linguistics/Linguistique de contact*. Berlin, New York: Walter de Gruyter.
Gorter, Durk 1997a: Dutch-West Frisian. In: Hans Goebl, Peter H. Nelde, Zdeněk Starý, Wolfgang Wölck (eds.) 1997, 1152–1158.
Gorter, Durk 1997b: Friezen als Europese taalminderheid [Frisians as a European language minority], in: Philippus H. Breuker, Antheun Janse (eds.) 1997, 286–292.
Gorter, Durk, Reitze J. Jonkman 1995: *Taal yn Fryslân op 'e nij besjoen* [Language in Friesland seen anew]. Ljouwert: Fryske Akademy.
Gorter, Durk, Gjalt H. Jelsma, Pieter H. van der Plank, K. de Vos 1984: *Taal yn Fryslân. Undersyk nei taalgedrach en taalhâlding yn Fryslân* [Language in Friesland: A study of language behavior and language attitude in Friesland]. Ljouwert: Fryske Akademy.
Grice, H(erbert) Paul 1975: Logic and conversation. In: Peter Cole, Jerry L. Morgan (eds.) 1975: *Syntax and Semantics*. Vol. 3: *Speech Acts*. New York: Academic Press, 41–58.
Grucza, Franciszek 1979: Rozwój i stan glottodydaktyki polskiej w latach 1946–1945 [The development and the state of Polish glottodidactics in the years 1946–1945]. In: Franciszek Grucza (ed.) 1979: *Polska myśl glottodydaktyczna 1945–1975. Wybór artykułów z zakresu glottodydaktyki ogólnej* [Polish glottodidactics in the years 1946–1945. A selection of articles from the domain of general glottodidactics]. Warszawa: Państwowe Wydawnictwo Naukowe, 5–16.
Grucza, Franciszek 1983a: *Zagadnienia metalingwistyki. Lingwistyka – jej przedmiot, lingwistyka stosowana* [Questions of metalinguistics: Linguistics – its subject, applied linguistics]. Warszawa: Państwowe Wydawnictwo Naukowe.
Grucza, Franciszek 1983b: Zum Gegenstand und zur inneren Gliederung der Linguistik und Glottodidaktik. *Kwartalnik Neofilologiczny* XXX (3), 217–234.
Grucza, Franciszek 1994: O wieloznaczności wyrazu „język", heterogeniczności związanych z nim desygnatów i istocie rzeczywistych języków ludzkich [On the ambiguity of the word "language", heterogeneity of its related designates and the essence of real human languages]. *Przegląd Glottodydaktyczny* 13, 7–37.
Haarmann, Harald 1988: Sprachen- und Sprachpolitik. In: Ulrich Ammon, Norbert Dittmar, Klaus J. Mattheier (eds.) 1988: *Sociolinguistics. Soziolinguistik*. Berlin: Walter de Gruyter, 1660–1678.
Haarmann, Harald 1989: Functional aspects of language varieties – A theoretical-methodological approach. In: Ulrich Ammon (ed.) 1989: *Status and Function of Languages and Language Varieties*. Berlin: Walter de Gruyter, 153–93.

Habermas, Jürgen 1987 [1981]: *Theorie des kommunikativen Handelns*. Bd. 1. *Handlungsrationalität und gesellschaftliche Rationalisierung*. Frankfurt am Main: Suhrkamp.
Haeckel, Ernst Heinrich 1988 [1866]: *Generelle Morphologie des Organismus*. Bd. 2: *Allgemeine Entwicklungsgeschichte*. Reprint. Berlin: Walter de Gruyter.
Halliday, M(ichael) A(lexander) K(irkwood) 1972 [1970]: Language structure and language function. In: John Lyons (ed.) 1972: *New Horizons in Linguistics*. Harmondsworth (etc.): Penguin Books, 140–165.
Halliday, M(ichael) A(lexander) K(irkwood) 1973: *Explorations in the Functions of Language*. London: Edward Arnold.
Harris, Roy 1990: The integrationist critique of orthodox linguistics. In: Michael P. Jordan (eds.) 1990: *The Sixteenth LACUS Forum, August 1989, Queen's University, Canada*. Chapel Hill, NC: The Linguistic Association of Canada and the United States, 63–77.
Haugen, Einar 1972: Language ecology. In: Anwar S. Dil (ed.) 1972: *The Ecology of Language. Essays by Einar Haugen*. Stanford, CA: Stanford University, 324–339.
Hawley, Amos H. 1950: *Human Ecology: A Theory of Community Structure*. New York: The Ronald Press.
Heinz, Adam 1978: *Dzieje językoznawstwa w zarysie* [A history of linguistics in outline]. Warszawa: Państwowe Wydawnictwo Naukowe.
Helbig, Gerhard 1973: *Geschichte der neueren Sprachwissenschaft. Unter dem besonderen Aspekt der Grammatik-Theorie*. Leipzig: Bibliographisches Institut.
Hempel, Carl G(eorg) 1965 [1959]: The logic of functional analysis. In: Carl G(eorg) Hempel 1965. *Aspects of Scientific Explanation. And Other Essays*. New York: The Free Press; London: Collier-Macmillan, 297–330 [(Reprint with alterations) In: Llewellyn Gross (ed.) 1959: *Symposium on Sociological Theory*. New York: Harper & Row, 271–307].
Horn, Klaus 1968: *Zweckrationalität in der modernen Architektur. Zur Ideologiekritik des Funktionalismus*. In: Heide Berndt, Alfred Lorenzer, Klaus Horn (eds.) 1968, 105–153.
Hornborg, Alf 1996: Ecology as semiotics: Outlines of a contextualist paradigm for human ecology. In: Philippe Descola, Gisli. Palsson (eds.) 1996: *Nature and Society: Anthropological Perspectives*. London: Routledge, 45–62.
Huntington, Samuel P. 1996: *The Clash of Civilizations and the Remaking of World Order*. New York: Simon & Schuster.
Ingold, Tim 1999 [1996]: Social relations, human ecology, and the evolution of culture: An exploration of concepts and definitions. In: Andrew Lock, Charles R. Peters (eds.) 1999: *Handbook of Human Symbolic Evolution*. Oxford: Blackwell, 178–203.
Ivić, Milka 1965 [1963]: *Trends in Linguistics*. Trans. Muriel Heppell. The Hague: Mouton (Janua linguarum. Series minor. Nr. 42) [*Pravci u lingvistici*. Ljubljana: Državna založba Slovenije].
Jakobson, Roman (Osipovič) 1960: Closing statement. Linguistics and poetics. In: Thomas A(lbert)Sebeok (ed.) 1960: *Style in Language*. Cambridge: The Massachusetts Institute of Technology Press, 350–377.
James, William 1890: *Principles of Psychology*. Vols. 1 and 2. New York: H. Holt.

Janse, Antheun 1997: Graaf Willem II van Holland en de Friese vrijheid [Count William II of Holland and the Frisian freedom]. In: Philippus H. Breuker, Antheun Janse (eds.) 1997, 77–86.

Jansma, Lammert Gosse 1997a: Onkerkelijkheid, orthodoxie en de regionale factor [Churchlessness, orthodoxy and the regional factor]. In: Philippus H. Breuker, Antheun Janse (eds.) 1997, 245–251.

Jansma, Lammert Gosse 1997b: Modernization, church membership, orthodoxy, and Frisian identity. In: Brunon Synak, Tomasz Wicherkiewicz (eds.) 1997, 245–255.

Jensma, Goffe (Th.) 1998: *Het rode tasje van Salverda. Burgerlijk bewustzijn en de Friese identiteit in de negentiende eeuw* [A red briefcase of Salverda. A bourgeois consciousness and the Frisian identity in the nineteenth century]. Ljouwert/Leeuwarden: Fryske Akademy.

Jonkman, Reitze Jehannes 1997: Op weg van onderschikking naar nevenschikking: Gewijzigde verhoudingen tussen het Fries en het Nederlands sinds de Franse Tijd [On the way from subordination to equality: Modified relations between Frisian and Dutch since the French time]. In: Philippus H. Breuker, Antheun Janse (eds.) 1997, 252–259.

Joseph, John J. 1995: Trends in twentieth-century linguistics: An overview. In: E(rnst) F(riderik) K(onrad) Koerner and R. E. Asher (eds.) 1995, 221–233.

Kelly, George A(lexander) 1955: *The Psychology of Personal Constructs*. New York: W. W. Norton.

Knapp, Mark L. 1978: *Nonverbal Communication in Human Interaction*. 2nd edition. New York: Holt-Rinehart and Winston.

Koerner, E(rnst) F(riderik) K(onrad), R. E. Asher (eds.) 1995: *Concise History of the Language Sciences. From Sumerians to the Cognitivists*. Oxford, New York, Tokyo: Elsevier.

Koerner, E(rnst). F(riderik) K(onrad), Aleksander Szwedek (eds.) 2001: *Towards a History of Linguistics in Poland. From the Early Beginnings to the End of the Twentieth Century* (Series III – Studies in the History of the Language Sciences. Volume 102). Amsterdam, Philadelphia: John Benjamins.

Koerner, E[rnst] F[riderik] Konrad 1993: The problem of metalanguage in linguistic historiography. *Studies in Language* 17 (1), 111–134.

Korżyk, Krzysztof 1999: Język i gramatyka w perspektywie 'komunikatywizmu' [Language and grammar in the perspective of 'communicativism']. In: Aleksy Awdiejew (ed.) 1999: *Gramatyka komunikacyjna* [The communicational grammar]. Warszawa, Kraków: Wydawnictwo Naukowe PWN, 9–32.

Kress, Gunther R. (ed.) 1976: *Halliday: System and Function in Language. Selected Papers*. London: Oxford University Press.

Kull, Kalevi 1998a: On semiosis, Umwelt, and semiosphere. *Semiotica* 120 (3/4), 299–310.

Kull, Kalevi 1998b: Semiotic ecology: Different natures in the semiosphere. *Sign Systems Studies* 26, 344–371.

Kull, Kalevi 2000: Organisms can be proud to have been their own designers. *Cybernetics and Human Knowing* 7 (1), 44–55.

Kull, Kalevi 2001: Ecosemiotics and the semiotics of nature. *Sign Systems Studies* 29 (1), 71–81.

Lamb, Sydney M(acdonald) 1966: *Outline of Stratificational Grammar*. Washington D.C.: Georgetown University.

Lambrecht, Knud 1994: *Information Structure and Sentence Form: Topic, Focus, and the Mental Representations of Discourse Referents*. Cambridge, England, New York, NY: Cambridge University Press (Cambridge Studies in Linguistics 71).

Landsberg, Piotr 2003: Koncepcja wielości cywilizacji Samuela Huntingtona wobec konstruktywizmu społecznego. In: Anna Pałubicka, Andrzej P. Kowalski (eds.) 2003: *Konstruktywizm w humanistyce* [Constructivism in human sciences]. Bydgoszcz: Oficyna Wydawnicza Epigram, 135–138.

Leech, Geoffrey (Neil) 1990 [1983]: *Principles of Pragmatics*. Seventh impression: London, New York: Longman.

Levinson, Stephen C. 2003 [1983]: *Pragmatics*. 15th printing. Cambridge: Cambridge University Press.

Lizis, Elżbieta (Magdalena) 1996: Badania nad modelem zewnętrznego opisu języka [Studing the model of the external description of language]. *Acta Universitatis Wratislaviensis* No 1840. *Studia Linguistica* XVII: *Heteronomie języka*, 1996: 27–44.

Lockwood, David 1972: *An Introduction to Stratificational Linguistics*. Harcourt Brace Jovanovitch.

Loos, Adolf 1981 /1964/ {1962 [1908]}: Ornament und Verbrechen. In: Ulrich Conrads 1981 /1964/: Programme und Manifeste zur Architektur des 20. Jahrhunderts. 2nd editon. Trans. Henni Korssakoff-Schröder. Braunschweig: Vieweg Verlag (Ullstein Bauwelt Fundamente I), 15–21/.{In: Adolf Loos 1962: Sämtliche Schriften in zwei Bänden. Edited by Franz Glück. Wien, München: Herold, 276–288 [Buchhandlung]}.

Lorenzer, Alfred 1968: Städtebau: Funktionalismus und Sozialmontage? Zur sozialpsychologischen Funktion der Architektur. In: Heide Berndt, Alfred Lorenzer, Klaus Horn (eds.) 1968, 51–104.

MacDonald, Graham, Philip Pettit 1981: *Semantics and Social Science*. London, Boston: Routledge and Kegan Paul.

Makkai, Adam 1993: *Ecolinguistics: Towards a New Paradigm for the Science of Language?* London, New York: Pinter Publishers – Budapest: Akadémiai Kiadó (Open Linguistics Series).

Malinowski, Bronisław (Kasper) 1922: *Argonauts of the Western Pacific*. London: Routledge & Paul Kegan.

Malinowski, Bronisław (Kasper) 1926: Anthropology. In: *Encylopeadia Britannica*. First Supplementary Volume, London, New York: The Encyclopaedia Britannica Inc., 132–133.

Malinowski, Bronisław (Kasper) 1949 [1923]: Supplement I. The problem of meaning in primitive languages. In: Charles K(ay) Ogden, Ivor A(rmstrong) Richards 1949: *The Meaning of Meaning. A Study of the Influence of a Language Upon Thought and of the Science of Symbolism*. With supplementary essays by Bronisław Malinowski (Kasper), Francis G(raham) Crookshank. London: Paul Kegan, Trench, Trubner and Company; New York: Harcourt, Brace and Company. 10th edition. London: Routledge & Paul Kegan, 296–336.

Maturana, Humberto R., Francisco J. Varela, 1987: *The Tree of Knowledge: The Biological Roots of Understanding*. Boston: New Science Library.

McLuhan, Marshall 1964: *Understanding Media: The Extensions of Man*. New York: Signer Books.
Merton, Robert K(ing) 1957 [1949]: Manifest and latent functions. In: Robert K(ing) Merton 1957 [1949]: *Social Theory and Social Structure*. Revised and enlarged edition. Glencoe, Illinois: The Free Press, 99–84.
Milewski, Tadeusz 1973 [1965]: *Introduction to the Study of Language*. Trans. Marsha Brochwicz. The Hague, Paris: Mouton [*Językoznawstwo*. Warszawa: Państwowe Wydawnictwo Naukowe (Polish Scientific Publishers)].
Milewski, Tadeusz 2004 [1965]: *Językoznawstwo* [Linguistics]. 7th altered edition. Wydawnictwo Naukowe PWN [Państwowe Wydawnictwo Naukowe].
Misiak, Małgorzata 2006: *Łemkowie. W kręgu badań nad mniejszościami etnolingwistycznymi w Europie* [The Lemkos. Within the scope of the research into ethnolinguistic minorities in Europe]. Wrocław: Wydawnictwo Uniwersytetu Wrocławskiego (Acta Universitatis Wratislaviensis No 2858).
Moffett, James 1983: *Teaching the Universe of Discourse*. Portsmouth: Heinemann.
Moffett, James 1987: *Detecting Growth in Language*. Portsmouth: Heinemann.
Molen, Sytse J. van der 1974: *Ta de Fryske folkskunde* II. *Samling losse stikken* [On the Frisian Ethnology. A Collection of Loose Fragments] Grins [Groningen]: Frysk Ynstitut oan de Ryksuniversiteit to Grins.
Moles, Abraham 1971: *Psychologie du Kitsch. L'art du bonheborn* Paris: Maison Mame & Munich: Calr Hauser Verlag.
Mukařovský, Jan 1970 [1936]: *Aesthetic Function. Norm and Value as Social Facts*. Trans. from Czech with notes and afterword by M. E. Suino. Ann Arbor: Michigan Slavic Contributions [*Estetická funkce. Norma a hodnota jako socialni fakty*. Praha: Borový].
Nagel, Ernst 1967 [1956]: A formalization of functionalism. In: Ernest Nagel, *Logic Without Metaphysics*. New York: The Free Press [Glencoe, Illinois: The Free Press], 247–287.
Nęcki, Zbigniew 2000: *Komunikacja międzyludzka* [Human communication]. Kraków: Oficyna Wydawnicza.
Nöth, Winfried 1998: Ecosemiotics. *Sign System Studies* 26, 332–343.
Nöth, Winfried 2001: Ecosemiotics and the semiotics of nature. *Sign Systems Studies* 29 (1), 71–81.
Oertel, Hanns 1901: *Lectures on the Study of Language*. New York, London: C. Scribner's sons (Yale bicentennial publications).
Ormrod, Jeanne Ellis 1995: *Educational Psychology: Principles and Applications*. Englewood Cliffs, NJ: Merrill; Toronto: Maxwell Macmillan Canada; New York: Maxwell Macmillan International.
Park, Robert E(zra), Ernest W(atson) Burgess 1921: *Introduction to the Science of Sociology*, including the original index to basic sociological concepts. Chicago, IL: University of Chicago Press.
Parsons, Talcott 1949: *The Structure of Social Action*. New York: The Free Press.
Parsons, Talcott, Edward A. Shils (eds.) 1967 [1951]: *Toward a General Theory of Action*. 6th printing. Cambridge, MA: Harvard University Press.

Parsons, Talcott, Edward A. Shils, Gordon W. Allport, Clyde Kluckhohn, Henry A. Murray, Jr., Robert R. Sears, Richard C. Sheldon, Samuel A. Stouffer, Edward C. Tolman 1967 [1951]: Some fundamental categories of the theory of action: A general statement. In: Talcott Parsons, Edward A. Shils (eds.) 1967, 3–29.

Piebenga, Jan Tjittes 1954: Sêgen fan kening Redbad [Legends about King Redbad] (On the basis of J[acobus] P[ieter] Wiersma, Friese Sagen). In: Jan Tjittes Piebenga 1954: *Frysk Lêsboek* [Frisian Reader] Part 2. Grins. Djakarta: J. B. Wolters, 110–117.

Pietraszko, Stanisław 1992: *Studia o kulturze* [Studies on culture]. Wrocław: Wydawnictwo Uniwersytetu Wrocławskiego.

Plank, Pieter H. van der 1987: Frisian language use and ethnic identity. *International Journal of the Sociology of Language* 64, 9–20.

Pobojewska, Aldona (Teresa) 1996: *Biologia i poznanie. Biologiczne „a priori" człowieka a realizm teoriopoznawczy* [Biology and cognition. Biological "apriori" of man and the theoretical-cognitive realism]. Łódź: Wydawnictwo Uniwersytetu Łódzkiego.

Pobojewska, Aldona (Teresa) 1998: *Istota żywa jako podmiot. Wybór pism Jakoba Johannesa von Uexkülla. Wprowadzenie, wybór pism i przypisy Aldona Pobojewska* [The living being as a subject. The selection of Jakob Johannes von Uexküll's writings. Introduction, selection and footnotes by Aldona Pobojewska]. Trans. Aldona Pobojewska, Małgorzata Półrola. Łódź: Studio Wydawnicze KARTA.

Podsiad, Antoni 2000: *Słownik terminów i pojęć filozoficznych* [A dictionary of philosophical terms and notions]. Warszawa: Instytut Wydawniczy Pax.

Popper, Karl 1972: *Objective Knowledge: An Evolutionary Approach*. Oxford: The Clarendon Press.

Radcliffe-Brown, Alfred Reginald 1922: *The Andaman Islanders*. New York: Free Press.

Reiter, Norbert (ed.) 1999: *Eurolinguistik ein Schritt in die Zukunft. Beiträge zum Symposium vom 24. bis 27. März 1997 im Jagdschloß Glienicke (bei Berlin)*. Berlin: Harrassowitz Verlag.

Reiter, Norbert 1986: Die Irrtümer um den Idiolekt. *Incontri Linguistici* 11, 137–151.

Rice, Frank A. (ed.) 1962: *Study of the Role of Second Languages in Asia, Africa and Latin America*. Washington D.C.: Center for Applied Linguistics of the Modern Language Association of America.

Riegler, Alexander 2001: Towards a radical constructivist understanding of science. *Foundations of Science, special issue on "The Impact of Radical Constructivism on Science"*, edited by Alexander Riegler, vol. 6, no. 1–3: 1–30.

Riegler, Alexander, Markus Peschl, Astrid von Stein (eds.) 1999: *Understanding Representation in the Cognitive Sciences: Does Representation Need Reality?* New York: Kluwer Academic/Plenum.

Robertson, Roland 1992: *Globalization: Social Theory and Global Culture*. London: Sage.

Robins, R(obert) H(enry) 1965 [1964]: *General Linguistics. An Introductory Survey*. Second impression London: Longmans.

Robins, R(obert) H(enry) 1974 [1967]: *A Short History of Linguistics*. 4th edition. Bloomington, London: Indiana University Press [London: Longmans].

Roblyer, Margaret D., Jack Edwards, Mary Anne Havriluk 1997: *Integrating Educational Technology into Teaching*. Upper Saddle River, N.J.: Prentice Hall.
Sadock, Jerrold M. 1974: *Toward a Linguistic Theory of Speech Acts*. New York: Academic Press.
Saussure, Ferdinand de 1959 [1916]: *Course in General Linguistics*. Trans. Wade Baskin. New York: Philosophical Library [*Cours de linguistique générale*. Publié par Charles Bally et Albert Sechehaye. Avec la collaboration de Albert Riedlinger. Lausanne & Paris: Payot].
Saussure, Ferdinand de 1983 {1972 [1916]}: *Course in General Linguistics*. From the critical French edition of Tulio De Mauro translated and annotated by Roy Harris London: Duckworth {*Cours de linguistique générale*. Publié par Charles Bally et Albert Sechehaye. Édition critique de Tullio De Mauro. Paris: Grande bibliothèque Payot}.
Saville-Troike, Muriel 1982: *The Ethnography of Communication. An Introduction*. Oxford, New York: Basil Blackwell.
Schaff, Adam 1962 [1960]: *Introduction to Semantics*. Oxford, New York: Pergamon Press [*Wstęp do semantyki*. Warszawa: Państwowe Wydawnictwo Naukowe].
Searle, John R(oger) 1969: *Speech Acts. An Essay in the Philosophy of Language*. Cambridge: Cambridge University Press.
Searle, John R(oger) 1992: *The Rediscovery of the Mind*. Cambridge, MA: The Massachusets Institute of Technology Press.
Seebold, Elmar 2001: Die Friesen im Zeugnis antiker und spätantiker Autoren. The Frisians in the writings of classical and post-classical authors. In: Horst Haider Munske (ed.) 2001: *Handbuch des Friesischen. Handbook of Frisian Studies*. Tubingen Max Niemeyer, 479–487.
Sebeok, Thomas A(lbert) (gen. ed.) 1986: *Encyclopedic Dictionary of Semiotics*. Berlin, New York, Amsterdam: Mouton de Gruyter (Approaches to Semiotics 73).
Siek-Piskozub, Teresa 2004: Social constructivism in foreign language education. *Acta Universitatis Nicolai Copernici. English Studies XIII. Humanities and Social Sciences*. Volume 362, 11–25.
Smoczyński, Wojciech 2001: Jerzy Kuryłowicz as Indo-Europeanist and theorist of language. In: E(rnst) F(rierik) K(onrad) Koerner, Aleksander Szwedek 2001, 255–271.
Sokołowska, Olga 2001: *A Cognitive Study of Speech Acts*. Gdańsk: Wydawnictwo Uniwersytetu Gdańskiego.
Sperber, Dan, Deirdre Wilson 1995 [1986]: *Relevance. Communication and Cognition*. 2nd edition. Oxford: Blackwell Publishing.
Stampe, Dennis W. 1975: Meaning and truth in the theory of speech acts. In: Peter Cole, Jerry L. Morgan (eds.) 1975: *Syntax and Semantics*. Vol. 3: *Speech Acts*. New York: Academic Press, 1–39.
Stewart, William A. 1962: An outline of linguistic typology for describing multilingualism. In: Frank A. Rice (ed.) 1962, 15–25.
Stewart, William A. 1970 [1968]: A sociolinguistic typology for describing national multilingualism. In: Joshua A. Fishman (ed.) 1970, 531–545.

Synak, Brunon, Tomasz Wicherkiewicz (eds.) 1997: *Language Minorities and Minority Languages in the Changing Europe. Proceedings of the 6th International Conference on Minority Languages. Gdańsk, 1–5 July, 1986.* Gdańsk: Wydawnictwo Uniwersytetu Gdańskiego.

Sztompka, Piotr 2002: *Socjologia. Analiza społeczeństwa* [Sociology. An analysis of society). Kraków: Społeczny Instytut Wydawniczy Znak.

Trubetzkoy, Nikolaj Sergeevič 1967 [1939]: *Grundzüge der Phonologie* [*Travaux du Cercle Linguistique de Prague* VII]. 4th edition. Göttingen: Vandenhoeck und Ruprecht.

Turowski, Jan 1993: *Socjologia. Małe struktury społeczne* [Sociology. Small social structures]. Lublin: Wydawnictwo Towarzystwa Naukowego Katolickiego Uniwersytetu Lubelskiego.

Varela, Francisco J., Evan Thomson, Eleanor Rosch 1991: *The Embodied Mind: Cognitive Science and Human Experience.* Cambridge, MA: The Massachusetts Institute of Technology Press.

Wąsik, Elżbieta (Magdalena) 1998: Przyczynek do modelu opisu ekologicznego języków mniejszościowych w Europie (na przykładzie języka fryzyjskiego) [A contribution to the model of an ecological description of minority languages in Europe on the basis of Frisian)]. In: Barbara Lewandowskia-Tomaszczyk, Agnieszka Leńko-Szymańska (eds), 1998: *I Kongres Neofilologiczny, Łódź, 5–7 listopada 1998 r. Materiały Konferencyjne* [I Neophilological Congress, Łódź, November 5–7, 1998. Conference Materials]. Łódź: Wydawnictwo Uniwersytetu Łódzkiego, 54–55.

Wąsik, Elżbieta (Magdalena) 1999a: Domänen und Aufgaben der externen Sprachbeschreibung. Am Beispiel des Friesischen. In: Norbert Reiter (ed.) 1999, 303–316.

Wąsik, Elżbieta (Magdalena) 1999b: *Ekologia języka fryzyjskiego. Z badań nad sytuacją mniejszości etnolingwistycznych w Europie* (The Ecology of Frisian. From the studies on the situation of ethnolinguistic minorities in Europe). Wrocław: Wydawnictwo Uniwersytetu Wrocławskiego (Studia Linguistica XIX. Acta Universitatis Wratislaviensis No 2118).

Wąsik, Elżbieta (Magdalena) 1999c: Przyczynek do modelu opisu ekologicznego języków mniejszościowych w Europie (na przykładzie języka fryzyjskiego) [A contribution to the model of an ecological description of minority languages In Europe on the basis of Frisian)]. *Kwartalnik Neofilologiczny* (Warszawa: Polska Akademia Nauk) XLI (1–2), 57–64.

Wąsik, Elżbieta (Magdalena) 1999d: Dziedziny użycia języka z perspektywy językoznawstwa zewnętrznego [Domains uf language use from the perspective of external linguistics]. *Rozprawy Komisji Językowej* (Wrocław: Wrocławskie Towarzystwo Naukowe) XXV, 49–57.

Wąsik, Elżbieta (Magdalena) 2000a: Towards redefining the concept of the ecology of language in the framework of human-centered linguistics (with special reference to Frisian-speaking linkages). In: Victor H(use) Yngve, Zdzisław Wąsik (eds.) 2000, 29–31.

Wąsik, Elżbieta (Magdalena) 2000b: O potrzebie nauczania języka ojczystego w warunkach dyglosji (na przykładzie języka fryzyjskiego). *Neofilolog. Czasopismo Polskiego Towarzystwa Neofilologicznego* 19 (Poznań), 22–28.

Wąsik, Elżbieta (Magdalena) 2001: Ethnic identity in a semiotic perspective (on the example of Frisian). In: Eero Tarasti (ed.) 2001: *ISI Congress Papers. Nordic-Baltic Summer Institute for Semiotic and Structural Studies. Part I. June 12–21, 2000 in Imatra, Finland. Plenary Lectures and Sections: Signs of Nation and History*. Imatra: International Semiotics Institute. Cultural Centre, 267–282.

Wąsik, Elżbieta (Magdalena) 2003: On the idea of an 'ecological grammar' of verbal discourse from a human-centered perspective. *Scripta Neophilologica Posnaniensia* V (Poznań: Wydział Neofilologii. Uniwersytet im. Adama Mickiewicza w Poznaniu), 263–274.

Wąsik, Elżbieta (Magdalena) 2004: Describing Frisian communities in terms of human linguistics. In: Victor H(use) Yngve, Zdzisław Wąsik (eds.) 2004, 214–224.

Wąsik, Elżbieta (Magdalena) 2005: Sociological pragmatics from a hard-science perspective. A side-note to the conception of human linguistics. *Scripta Neophilologica Posnaniensia* VII (Poznań: Wydział Neofilologii Uniwersytet im. Adama Mickiewicza w Poznaniu), 181–193.

Wąsik, Elżbieta (Magdalena) 2006: Linguistic properties of communicating individuals in the construction of intersubjective world of meanings. In: Edyta Lorek-Jezińska, Teresa Siek-Piskozub, Katarzyna Więckowska (eds.) 2006: *Worlds in the Making: Constructivism and Postmodern Knowledge*. Toruń: Nicolaus Copernicus University, 87–101.

Wąsik, Elżbieta (Magdalena) 2007: *Język – narzędzie czy właściwość człowieka? Założenia gramatyki ekologicznej lingwistycznych związków międzyludzkich* [Language – a tool or property of man? Towards an idea of ecological grammar of human linkages]. Poznań: Wydawnictwo Naukowe im. Adama Mickiewicza w Poznaniu (Seria Filologia Angielska nr 28).

Wąsik, Elżbieta (Magdalena), Zdzisław Wąsik 2004: "Ecological grammar" of communicative linkages: Between bioinvasion and cultural transmission in the globalized world. In: Anna Duszak, Urszula Okulska (eds.) 2004: *Speaking from the Margin: Global English from a European Perspective*. Frankfurt am Main et al.: Peter Lang (Polish Studies in English Language and Literature 11), 133–144.

Wąsik, Zdzisław 1986: W sprawie koncepcji antropocentrycznej języka na potrzeby lingwistyki stosowanej. Na marginesie książki Franciszka Gruczy pt. Zagadnienia metalingwistyki. *Acta Universitatis Wratislaviensis* No 905. *Studia Linguistica* X (Wrocław: Wydawnictwo Uniwersytetu Wrocławskiego), 91–102.

Wąsik, Zdzisław 1993a: Z zagadnień ekologicznej charakterystyki języków mniejszościowych na przykładzie dialektu Arumunów na Bałkanach. In: Ilona Czamańska, Leszek Mroziewicz, Włodzimierz Pająkowski (eds.) 1993: *Narody bałkańskie XVI–XX wieku*. Poznań: Uniwersytet Adama Mickiewicza w Poznaniu (Balcanica Posnaniensia. Acta et studia VI), 355–366.

Wąsik, Zdzisław 1993b: O pojęciu ekologii języka – tytułem wstępu [On the notion of the ecology of language – by the way of an introduction]. In: Zdzisław Wąsik (ed.) 1993: *Z zagadnień ekologii języka* [From the issues of the ecology of language]. Wrocław: Wydawnictwo Uniwersytetu Wrocławskiego (Acta Universitatis Wratislaviensis No 1455. Studia Linguistica XVI), 13–23.

Wąsik, Zdzisław 1997: *Systemowe i ekologiczne właściwości języka w interdyscyplinarnych podejściach badawczych* [Systemic and ecological properties of language in interdisciplinary investigative approaches]. Wrocław: Wydawnictwo Uniwersytetu Wrocławskiego (Acta Universitatis Wratislaviensis No 1948. Studia Linguistica XVIII).

Wąsik, Zdzisław 1997b: Interdisciplinary Domains in the Studies of Minority Languages and the Division of Linguistic Labor. In: Brunon Synak, Tomasz Wicherkiewicz (eds.) 1997, 79–91.

Wąsik, Zdzisław 1999: Sprachautonomie, Sprachvariabilität und das Problem der Mehrsprachigkeit aus der Perspektive der komparativen Linguistik. In: Norbert Reiter (ed.) 1999, 57–71.

Wąsik, Zdzisław 2000: On the heteronomous nature of language and its autonomization from the properties of communicating individuals and linkages. In: Victor H(use Yngve), Zdzisław Wąsik (eds.) 2000, 31–32.

Wąsik, Zdzisław 2001a: On the biological concept of subjective significance: A link between the semiotics of nature and the semiotics of culture. *Sign Systems Studies* 29 (1): 83–106.

Wąsik, Zdzisław 2001b: The development of general linguistics in the history of the language sciences in Poland. Late 1860s – Late 1960s. In: E(rnst), F(riderik) K(onrad) Koerner, Aleksander Szwedek 2001, 3–51.

Wąsik, Zdzisław 2003: *Epistemological Perspectives on Linguistic Semiotics.* Frankfurt am Main et al: Peter Lang (Polish Studies in English Language and Literature 8).

Wąsik, Zdzisław 2006a: Understanding the existence modes of language and the division of linguistic labor. In: Anna Duszak, Urszula Okulska (eds.) 2006: *Bridges and Barriers in Metalinguistic Discourse.* Frankfurt am Main, Berlin, Bern, Bruxelles, New York, Oxford, Wien: Peter Lang (Polish Studies in English Language and Literature 17), 43–56.

Wąsik, Zdzisław 2006b: Investigative perspectives in the construction of scientific reality: An epistemological outlook on the foundations of linguistic semiotics. In: Edyta Lorek-Jezińska, Teresa Siek-Piskozub, Katarzyna Więckowska (eds.) 2006: *Worlds in the Making: Constructivism and Postmodern Knowledge.* Toruń: Nicolaus Copernicus University, 21–35.

Wąsik, Zdzisław, Elżbieta (Magdalena) Wąsik, Maciej Kielar 2006: Applying Victor H. Yngve's theory of human linguistics to the analysis of interpersonal relationships among communication participants in social reality and literary fiction. In: Katarzyna Dziubalska-Kołaczyk (ed.) 2006: *IFAtuation: A Life in IFA. A Festschrift for Professor Jacek Fisiak on the Occasion of His 70th Birthday by His IFAtuated Staff from the School of English, AMU, Poznań.* Poznań: Wydawnictwo Naukowe im. Adama Mickiewicza, 709–728.

Wawrzyniak, Zdzisław 1974: Sposoby istnienia języka. *Biuletyn Polskiego Towarzystwa Językoznawczego* XXXII, 87–91.

Wierzbicka, Anna 1972: *Semantic Primitives.* Frankfurt: Athenäum.

Wirrer, Jan 1997: Scenarios of Endangeredness: Endangered Languages, Less Endangered Languages, Non-Endangered Languages. In: Brunon Synak, Tomasz Wicherkiewicz (eds.) 1997, 153–166.

Woodfield, Andrew 1976: *Teleology*. Cambridge, New York, Melbourne: Cambridge University Press.
Wright, Larry 1976: *Teleological Explanations. An Etiological Analysis of Goals and Functions*. Berkeley, Los Angeles, London: University of California Press.
Yngve, Victor H(use) 1969: On achieving agreement in linguistics. In: Robert I. Binnick et al. (eds.) 1969: *Papers from the Fifth Regional Meeting of the Chicago Linguistic Society, April 18–19, 1969*. Chicago: University of Chicago, Department of Linguistics, 455–462.
Yngve, Victor H(use) 1975a: The dilemma of contemporary linguistics. In: Adam Makkai, Valerie Becker Makkai (eds.) 1975: *The First LACUS Forum 1974*. Columbia, SC: Hornbeam Press, 1–16.
Yngve, Victor H(use) 1975b: Human linguistics and face-to-face interaction. In: Adam Kendon et al. (eds.) 1975: *Organization of Behavior in Face-to-Face Interaction*. The Hague: Mouton, 47–62 (Reprint in: Thomas R. Williams (ed.) 1975: *Socialization and Communication in Primary Groups*. The Hague: Mouton, 313–328).
Yngve, Victor H(use) 1975c: Toward a human linguistics. In: Robin L. Grossman, James San, Timothy J. Vance (eds.) 1975: *Papers from the Parasession on Functionalism*. Chicago: Chicago Linguistic Society, 540–555.
Yngve, Victor H(use) 1986: *Linguistics as a Science*, Bloomington, IN: Indiana University Press.
Yngve, Victor H(use) 1996: *From Grammar to Science: New Foundations for General Linguistics*. Amsterdam, Philadelphia: John Benjamins.
Yngve, Victor H(use) 2004a: Issues in hard-science linguistics. In: Victor H(use) Yngve, Zdzisław Wąsik (eds.) 2004, 14–26.
Yngve, Victor H(use) 2004b: An introduction to hard-science linguistics. In: Victor H(use) Yngve, Zdzisław Wąsik (eds.) 2004, 27–34.
Yngve, Victor H(use) 2004c: The conduct of hard-science research. In: Victor H(use) Yngve, Zdzisław Wąsik (eds.) 2004, 332–341.
Yngve, Victor H(use), Zdzisław Wąsik (eds.) 2000: Workshop: Exploring the Domain of Human-Centered Linguistics from a Hard-Science Perspective. *Societas Linguistica Europaea 33rd Annual Meeting: Naturalness and markedness in synchrony and diachrony*. Poznań: The School of English, Adam Mickiewicz University, Motivex.
Yngve, Victor H(use), Zdzisław Wąsik (eds.) 2004: *Hard-Science Linguistics*. London, New York: Continuum Books.
Yngve, Victor H(use), Zdzisław Wąsik 2004a: The riches of the new world. In: Victor H(use) Yngve, Zdzisław Wąsik (eds.) 2004, 321–325.
Yngve, Victor H(use), Zdzisław Wąsik 2004b: Coping with cultural differences. In: Victor H(use) Yngve, Zdzisław Wąsik (eds.) 2004, 326–331.
Zawadowski, Leon 1961: Fundamental relations in language contact. *Biuletyn Polskiego Towarzystwa Językoznawczego* 20, 3–26.
Zawadowski, Leon 1966: *Lingwistyczna teoria języka* [A linguistic theory of Language]. Warszawa: Państwowe Wydawnictwo Naukowe.
Zawadowski, Leon 1975: *Inductive Semantics and Syntax: Foundations of Empirical Linguistics*. The Hague: Mouton (Janua Linguarum. Series Major 58).

Index

Names and terms from the main text

act-/s, communicational 16, 67, 107, 109, 145, of communication 109, of communicational conduct 82, of comprehension and interpretation 70, 113, of speaking and understanding 39, of speech 15, 27, of thinking 32, speech 25–26, 46–47, 52, 69, 93, 95, 100, 113
Adamska-Sałaciak, Arleta (b. 1957) 18
Ammon, Ulrich 49
Angell, James R(owland) (1869–1949) 33
Antas, Jolanta (b. 1954) 97
anthropocentric, conception of language 56
anthropocentrism 16, 23
anthropological, linguistics 24–25, 70, properties 77, thought 102
anthropologism, cultural 24
anthropology 24, 68, cultural 26, 29–30, 33, 106, linguistic 26, 70, socio-cultural 35, 36
anthropomorphic, approach 32
anthroposemiotics 80
arbitrariness, and conventionality of linguistic signs 22
Austin, John L(angshaw) (1911–1960) 95
automatization 21
autonomization 78
autonomous, agent-/s 79, 83, 85, entities 110, (linkage-/s) 83, 85, organism-/s 73 phenomen-on/a 67, 108, system-/s 21, 32
autonomy, communicational 116, language 18, principle 5, 21, 22
Awdiejew, Aleksy (b. 1940) 24

Bańczerowski, Jerzy (b. 1938) 86
Bańka, Józef (Marian) (b. 1934) 70
Barker, Roger G(arlock) (1903–1990) 81
Bartoli, Matteo Giulio (1873–1946) 20
Bateson, Gregory (1904–1980) 81
Bednarczuk, Leszek (b. 1936) 10
Berger, Peter Ludwig (b. 1929) 58–60
Berndt, Heide (1938–2003) 31
Bertoni, Giulio (1878–1942) 20
bilingual, being 94, someone is 64, or multilingual 76, 111, out of necessity 119, linkages 141
bilingualism 24, 78, 122
bio- and anthroposemiotics 80
bioinvasion-/s 6, 12, 86, 89–90
biological, *a priori* of man 68, (conditioning-/s) 110, 113, and cultural diversity 26, 87, concept of "embodied semantics" 88, concept of language 19, conception of the linguistic ecosystem 111, conceptions of Humberto R. Maturana, Francisco J. Varela, Evan Thomson, and Eleanor Rosch 25, factors 59, features 70, nature of humankind 100, organicism 34, organism-/s 19, 25, 35, 65, 70, phenotypic features 88, roots of understanding 70, sciences 25, 33, sciences 33, tool 32
biologist-/s 68, 70, 75
biology 29–30, 32–33, 35–36, 66, 68, 93, 106, 145, philosophy of 49, 68
Beauvais, Vincent de (Vincentius Bellovacensis) Vincent of Beauvais (c. 1190–1264?) 140
Bois, Joseph Samuel 100

Boissevain, Jeremy (b. 1928) 83
Bopp, Franz (1791–1867) 16
Brentano, Franz (Clemens) (1838–1917) 33
Breuker, Philippus H. (b. 1939) 124–125, 132, 134, 137–138
Brugmann, Karl (1849–1919) 18
Bruno 125
Bühler, Karl (Ludwig) (1879–1963) 22, 40–43, 45–47
Burgess, Ernest W(atson) (1886–1966) 75
Carnap, Rudolf (1891–1970) 102
Carr, Harvey A. (1873–1954) 33
Cartesius (zob. Descartes, René) 102
Cassirer, Ernst (1874–1945) 101
change-/s, and evolution 22, (groupings of linkages) always 95, (teleological and etiological) explanation-/s of 5, 18–19, 32, in (of) the (linguistic) properties (of speakers and hearers) 95–96, 108, 110, in communicating individuals 68, 108, in ecological conditionings 111, in memberships 84, in the meaning of words 20, in the mind 21, language 17–19, 21–22, natural 32, possibility of 85, societal/social 128, 132, within an (a given) ecosystem 73, 79
changeability 21, 83
changeable, nature of linguistic structures 28, nature of multilingual linkages 145, linguistic knowledge 56, 109, practices 74, 80, 84, 111, communicational practice 112, (biological, cultural and social) conditionings 101, 113
channel-/s, (of) communication 42, 44, 51, 65–66, 74, 78, 80, 93–95, 99, 138–139, 144
Charles the Great (Charles le Magne, also: Charlemagne, c. 742–814), 124, 126, 139–140
Charles, the king of the Franks (Fri. *Karel fan Frankelân*) 126–127
chemistry 66, 93, 145

Chomsky, Noam A(vram) (b. 1928) 23
Christy, Thomas Craig (b. 1952) 18
Collinge, N(eville) E(dgar) (Oscar) (b. 1921) 17–18, 20
communicating, activities 51, 62, 65, agents 48, 112, at least in a couple 65, groups 80, individual-/s 6, 12, 15, 27, 45, 50–51, 55–56, 60–62, 64–68, 73–75, 79–83, 85–87, 90–92, 94–95, 99–101, 103, 107–111, 139, 142, 144–145, minds 46, people 39, 74, 82, 86, 88, 91–92, purposes 106, subject-/s 16, 107, with others 59
communication, acts of 109, among government functionaries 49, among their members 85, at home 130, by voice 60, channels (of) 51, 78, 80, conditionings of 26, constituents (of) 6, 42, education and 89, effectiveness of 97, effective 109, ethnography of 25, everyday 49, extra-semiotic reality of 43, genres of 52, global network of 86, group 55–56, 67, 74, 80, 111, 119, 145, human 6, 13, 15, 28, 58, 62, 66, 78–79, 88, 92, 102, 108, 145, in the 130, 144, international 25, interpersonal 16, 21, 26, 45, 57, 65, 69, 76–77, 88, 92, 96, 100, 112, 129, language (of) 77–78, 100, 115, 130, 141, 143, levels 116, linguistic 25, 58, 64, 69, 79, 87, 93, 99, 107, local 141, mass 78, means of 38, 50–51, 53, 68, 73, 78 –79, 117–119, 127, 133, national or international 120, nonverbal 28, 98, participant-/s (of) 12–13, 25, 28, 42, 44, 51, 55–56, 61–64, 66–67, 69, 71, 79, 82, 87, 93–94, 96, 99, 101, 107–108, 110–111, 141, 145, person-to-person 139, public and mass 78, 134, scheme 5, 38, 40, 42, sciences 74, 83, 85, 105, situations and domains of 120, social 21, 57, 92, 107, speech genres of 52, spoken

49, spoken and written 25, systems 77, theory 101, tool of 5, 29, form-/s of 58, 74, 76, 80, verbal 45, 63–64, 70–71, 79, 82, 92, 96, 98, 100, 107, 111, verbal and nonverbal 28, wider 49, world-wide-web 86

communicational, activit-y/ies 56, 59, 108, 138, acts 16, 67, 107, autonomy 116, capabilities and the aptitudes 56, constituents 66, contexts 65, disciplines 79, domain 63, event-/s 109, 111, forms of interpersonal linkage systems 79, function 107, goal-/s 95, 97, grammar 24, 80, groupings 110, linkages 6, 13, 108, 110, means 137, nature 51, needs 68, phenomena 80, practice-/s 109, 112, pragmatism 68–69, 113, processes 84, 111, properties of man 15, 67, 145, purposes 71, requirements 42, semiotic systems 115, settings 48, 74, 80, skills 109, strategies 97, studies 81, task-/s 28, 52, 55, 64, 66, 69, 76, 82, 89, 92, 94, 108, 113, 138

communicative, actions 68, acts 26, aptitudes 27, behavior 74, 79, 93–95, 112, collectivities 6, 55, communit-y/ies 10, 15, 60, 65–66, 69–70, 83, 86, domains of life 86, event-/s 25–26, 63, factors 86, faculties 51, function 22, 40–41, 47, grammar 24, impact 98, interactions 55, -interactive entities 85, linkage-/s 12–13, 70, 82–83, 109–112, means 5, 21, 51, minority 138, properties 63, 83, 85, relevance 94, situations 26, 130–131, 136, 138, values 63, 107, 109

communit-y/ies, communicative 10, 15, 60, 65–66, 69–70, 83, 86, confessional 77, discursive 26, 74–75, 100, English 86, ethnic 118, 132, existing 18, Frisian (speaking) 12, 28, 119, group 26, 85, 107, 117, homogeneous 18, 23, 55, 58, 83, 85, 113, individuals and 34, 73, 77, institutional 26, linguistic 13, 26, 51, 56, 58–59, 68, 79, 86–87, 90–91, 95, 107–108, 110–113, 117, 137, local 87, long-lasting 69, 113, national 26, of animals, plants, and microorganisms 89, 115, of West Frisians 138, particular 45, professional and/or confessional 77, small-group 26, 85, 117, speech 24, 49, 76–77, 85, 112, 121, 124, 135–136, static 83, task-realizing 5, 29, various 86

comparativism 16–17

concrete, activities 50, and individual 112, communicative situations 130–131, 136, constituents 40, data 65, existence mode 13, interactions 109, manifestations 75, meaning bearers 50, objects 57, observable whole 69, 94, people 92, situation 64, speech 21, speech act 69, 93, substance 30, textual data 17, utterances 64

construct-/s 93, assumed 95, human 62, personal 6, 57–58, 61, 69, pure 16, social 59, theoretical 65, 91, 100

construction-/s 6, 12, 18, 56, 62, social 6, 57–58, 112, mental 61

constructivism, as an investigative perspective 57, 61, as a set of theories 61, mentalist 25, social 57, 61, 63, 68, 113, sociological 58, radical 62, urbanist 30–31

constructivist-/s 61, 67, 69, cognitive 61, perspective 56, position 62, radical 62, social 59, 61, 69, theories 6, 56–57, 61

contact-/s (between/with individuals and linkages, among people, with reality) 17, 58, 66, 109, 131 (as constituent of communication scheme) 42, conversational 43, immediate 60, 102, internal and external 82, language (between languages) 20, 76, 109, linguistics 87, network of

112, physical 42, situations 76, 78, 95, 107, 116, varieties 78
contextualism, functionalist 24
conventionality, human 89, of linguistic signs 22
conversation-/s 25, 49, 63, 71, 94, 131, 142
conversational, analysis 24, contact 43, rules 96, implicature 97, language 131, situations 97
Cooley, Charles Horton (1864–1929) 58
creativity 21, 89
Creole, studies 24
cultural, and historical comparative aspect 20, anthropologism 24, anthropology 26, 29–30, 33, 35–36, artifacts 34–35, 50, cognitive sciences 25, context 24, 101, differences 28, diversity 26, ecosystems 83, 86, environment-/s 13, 55, 69, 88, 107, 122, forces 76, forms 34, goods 91, historical and personal experiences 25, knowledge 17, morphology 20, object 50, pattern-/s 34, physical and personal conditionings 26, products and behavior 34, relations 25, settings 73, symbols 7, 132, transmission 12
Danesi, Marcel (b. 1946) 87
Darwin, Charles Robert (1809–1882) 32
Dawkins, Richard (b. 1941) 80, 89
Deely, John (N.) (b. 1942) 103
Descartes, René (1596–1650) 102–103
destination-/s 26, 36
Devetak, Silvo 123
DeVito, Joseph A. (b. 1938) 26, 100
Dewey, John (1859–1952) 33
Di Cristo, Albert 78, 81, 82
Dijkstra, Waling Gerrits (1821–1914). 129
Dik, Simon C. (b. 1940) 47
discourse-/s, analysis 25, context of 48, functions 48, genres 79, organizational functions 46, patterns of (interpreted) 45, 74, 80, pragmatics 47, sociological 123, specific 48, verbal 12, 28, 81–82, 111
discursive, function 48, communities 26, 74–75, 100, ecosystems 79, patterns 111, practices 25, 87, 111
domain-/s of, (language) communication 120, 143, control 82, epistemology 13, Frisian (studies, language) 12, 116, 120, 130, Frisians and their language 10, individual experience 62, human-centered linguistics 12, investigated objects 21, language sciences 13, language use 6, 11, 48, 76, life 50, 86, public life 130, 142, North-Frisian dialects 121, occurrence, functions and social stratification 122, reference 70, 88, 92, 99, 101, social life 107, 109, 119, study 24, subjects as organisms 102, synchronic studies 120, the geography of dialectal distributions 19, the linguistic and non-linguistic theories 15, the methodology of linguistics 10, the neighboring disciplines 49, 76, the physiology of speech 19, the sciences of language 9, the sociology of language 122
domain-/s, and functions of language use 116, and tasks of external description of languages 11, cognitive 62, connected with the ecology of language 81, investigative 9–10, 15, 25, 29, 32, 47, 105, 121, logical 66–67, 69, 88, 91–93, 101, logical and physical 6, 92, physical 6, 16, 66, 69, 91–93, 100, 101, 145
Doroszewski, Witold (1899–1976) 20
Downes, William 22, 27, 83–84
Duranti, Allessandro (b. 1950) 26, 70
Durkheim, Emile (1858–1917) 59
ecolinkages 83, communicative 85
ecological, adaptiveness 38, and societal-cultural symbols 7, 132, conditionings 13, 82, 85, 88, 110–111, description 11, 76, 116, environments 110, factors 87, 110, gram-

mar-/s 5, 6, 9, 12, 74, 79–82, 86, 90, 110–112, linkages 7, 137, model 7, 11, 115, properties 79, 81, 115, -relational properties 73, 76, -relational variables 115, settings 87, situation 116, 122, specimen 73, status 7, 9, 116, studies 7, 10, 120, thought 81, variables 11, 76–77, 82,–116, view 13, way of reasoning 73, 75
ecologically determined, linkage 83, 85, 111, dynamic systems 6, 82
ecology 32, human, 73, 75, 79, linguistic 111, metalinguistic 77, 115, notion of 6, 13, 74, 79, of discursive communities 74, of Frisian 11–12, of language bearers 77, 115, of language communication 77–78, 115, of language-/s, 5, 9–12, 73–74, 76–79, 81, 115, of man 73–74, of man and society 79, of sign 80, of verbal discourse 82, 110, term 73, 75–76, 78
ecosystem-/s, linguistic 6, 80, 86, 111, language 73 –74, 78–79, societal 74, 79, of communicating people 74, 86, discursive 79, 83, 85–86, 89, human 75, 89–90, natural 89, natural and cultural 83, 86, cultural-natural 110
Edelman, Gerard M. (b. 1929) 89
Edwards, Jack (b. 1945) 61
Einstein, Albert (1879–1955) 103
embodied, "in the lived histories of organisms" 88, in the native language of Frisians 136, mind 70, semantics 24–25, 70, 88
Emmeche, Claus (b. 1956) 89–90
enactionism, as opposed to mentalist constructivism 25, investigative perspective of 70
enactive, concept of meaning 70, ecologism 68–69, 113
energy, amount of 99, 108, associated with gestures 65, 94, exchanges of 66, flow-/s 65, 67, 91, 94, 99, in the physical and chemical processes 32, means of 65–66, of vocal sound waves 65, other kinds of 28, representation of 65
Enninger, (Heinz Josef) Werner 79
environmental, conditionings 9, 16, 100, 110, dependencies 121, urbanist constructivism 31, factors 73, 79, 97, semiosis 12, settings 75
Erickson, Frederick 123
ethnic, character 136, community 26, 118, 132, consciousness 118, ecosystems 83, exponents 135, group-/s 118, 124–125, 127, 130–131, groupings 74, identity 11, 77, 116, 123–124, 130, 135–136, 138–139, 144, minority 117, 123, origin 131, solidarity 76–77, 116, 119, 136, separateness 7, 125, 134
ethnicity 7, 51, 118, 123–124, 129–130, 135–136, 144
etiological, approach 105, cause-and-effect-oriented 106, explanations 5, 32, (function) 41, 106, sense 34, terms 38
etiology 35, 37, 105
evolution, 17–18, 22, 75, biological 89, (of) language-/s 17–18, uniformitarian 19
evolutionary, dynamics 90, ladder 19, process 89
evolutionism 17, 32
existence, and continuity 33, domains 45, forms of 32, 46, forms of language 15, 19, 23, Frisian 141, mode-/s 6, 13, 51, 73, of a given element 105, of groups of people 84, of a universal language 62, of language 16, 109, of social linkages 44, of linkage systems 55, of the same domain of reference 70, of the world 62, of various forms of interactions 74, the Frisian ideology 124
external, causers 106, characteristics 7, 115–116, conditionings 65, 73, 76, 121, contacts 82, description 11,

and internal dimension 9, environments 58, factors 76, 87, 110, facts 120, history 77, 115, linguistics 10–11, 77, 115, 120, observers 48, 67–68, 112, observations 108, properties 120, purpose 106, reality 62, world 68
Fasold, Ralph W. (b. 1940) 48–49
Fawcett, Robin P. (b. 1937) 6, 45–47
Ferguson, Charles A(lbert) (b. 1921) 6, 48–49
Fill, Alwin 81
Firth, John R(upert) (1890–1960) 24, 46–47
Fishman, Joshua A(aron) (b. 1926) 48
Foley, William A. 25, 70, 88
Frings, Theodor (1886–1968) 20
Friso 125, 140
function-/s, a survey of 6, 45, 64, and dysfunctions 34, autotelic 44, Bühler's 45, basic 53, 55, cognitive 43, communication 26, communicational 107, communicative 22, 40, 42, 47–48, definition of 5, 39, discourse 48, discourse organizational 45, discursive 48, distinctive 22, emotive 42–43, esthetic 44, expressive 41, 45, for the sake of man 23, fulfillment of a 30, 106, Halliday's 45–47, ideational 45, impressive 40–43, 45, in context 46, in the teleological sense 34, intersubjective 45, intersubjective 6, Jakobson's 42, language 5, 38–39, 40–47, 52, 113, linguistic 46, 48, 50, Malinowski's functions 47, manifest or latent 34, meaning, information, and 48, mental 33, metalingual 42–43, natural 49, notion of 29, 32, 37–38, 46, 105–106, of a given element 49, of a particular usage 33, of human speech 41, 47, of language 6, 23, 39–40, 42, 45, 48–51, 105, 107, 113, of language use 78, 116, of linguistic elements 22, of speech acts 95, of the linguistic sign 40–41, 45, of the manifestation forms of language 50, of what is adequate 31, phatic 42–44, poetic 43, poetic 44, principal 106, primary 30, 53, psychological 33, referential 43, reflections on 105, representational 6, 40, 42, search for 36, searching for 22, semantic 22, 27, 40, 42–43, semantic-representational 43, 45, significative 86, social 33–34, 48–50, 143, symbolic 43, 45, 136, term 29, 35–36, 38–39, 49–50, 52, 105–106, textual 45, threefold 43, types of 37, 46, understanding-/s of 29, 35, 37–38, 45, within a given organism 36
functional, alternatives 34, analysis 34, approach 24, 30, 34, 106, characteristics 121, circles 32, components (of grammar) 46–47, consequences 34, 37, description 38, devise 47, equivalents 34, explanation 32, 39, 106, features 41, functional for one social group and dysfunctional for the other 34, grammar 47, invariance 29, invariants 30, linguistics 56, nature of language 107, objects 29, 105, organization of natural languages 47, outlook 106, properties 27, 40, 64, psychology 33, realms 38, 47–48, reasons 66, relations 38, 47, relationship 42, requirements 34, sphere 50, statements 29, 32, 35, 37, studies 35, styles 78, substitutes 34, thinking 30, 105–106, view 5, 29, 38, 47
functionalism, as an investigative perspective 5, 35, 38, essence of 107, framework of 34, cognitive view of 46, linguistic 29, 50, instrumentalist 30–31, 106, organicist 30, 106, psychologist 32–33, structuralist 22
functionalist, contextualism 24, instrumentalism 30, perspectives 29–30
functionally, relevant 22, 41, 106–107

Names and terms from the main text 169

Furdal, Antoni (b. 1928) 9, 10, 42, 44
Galilei, Galileo (1564–1642) 102
Gardiner, (Sir) Alan H(enderson) (1879–1963) 5, 39–40
Garvin, Paul L(ucian) (1919–1994) 52
Gawroński, Andrzej (1885–1927) 20, 100
Giddens, Anthony (b. 1938) 87
Galliéron, Jules (1854–1926) 19
Givón, Talmy (b. 1936) 47–48
Glasersfeld, Ernst von 61–63
global, and local 86, communication 86, English 86, language 86, linkage aggregations 6, 73, local and 108, message 98, network 86, scale 86–90, 110
globalization, and bioinvasion 6, 86, as a process 89, in human ecosystems 90, of international communication 25, on the cultural level 89, term 87
glottodidactics 27, 69
goal-/s, intended 26, S's 37, intentions or 69, 96 or task 97, communicative 97, of an organism 32, 35, 49, of individuals 95, of speech acts 46, -directed 32, 36, 45, 49, -oriented 41, 105–106
Gondebald (473–516) 126
Gorter, Durk 123, 128, 130–132, 134, 137–138, 141–142
grammar-/s, as a network of interpersonal and intersubjective linkages 7, 105, ecological 5–6, 9, 12, 74, 79–82, 86, 90, 110–112, of communicative linkages 12, of human linkages 12, of linguistic-linkages 5–6, 9, 79, of verbal discourse 12
Grice, H(erbert) Paul (1913–1988) 97
Grimm, Jakob (Ludwig Karl) (1785–1863) 16
Gropius, Walter (1883–1969) 31
Grucza, Franciszek (Krzysztof) (b. 1937) 6, 27, 56, 63–64, 75
Grutte Pier (Pier Gerlofs Donia, (c. 1480–1520) 129

Grutte Weird 129
Haarmann, Harald (b. 1946) 82
Habermas, Jürgen (b. 1929) 59–60
Haeckel, Ernst (Heinrich Philipp August) (1834–1919) 75
Halbertsma, Eeltsje Hiddes (1797–1858) 135
Halbertsma, Joost Hiddes (1789–1869) 133
Halliday, M(ichael) A(lexander) K(irkwood) (b. 1925) 6, 23–24, 45–47
Harris, Roy (b. 1931) 24
Harris, Thomas 26
Haugen, Einar (Ingvald) (1906–1994) 10, 76
Havriluk, Mary Anne 61
Hawley, Amos H(enry) (b. 1910) 75
hearer-/s 27, 42, 56, 63–64, 69, 75, 92–93, 95, 109
Heinz, Adam (1914–1984) 16–17, 20
Helbig, Gerhard (b. 1929) 20
Hempel, Carl G(ustav) (b. 1905) 34–35, 49
homocentric, and integrationist 5, 24, conception 24
homolinguistics 109
Horn, Klaus (b. 1934) 31
human-centered, approach 27, 68, 108, disciplines 81, framework 74, 79, 82, linguistics 12, 28, 80, 101, 107, 108–109, 111–112, 137, perspective 12–13, 15, 51, 81–82, 101, 110–111, pragmatics 91, view 28, 75
Hume, David (1711–1776) 103
Huntington, Samuel P. (1927–2008) 56
individual-/s, citizens 128, communicating 6, 12, 15, 19–22, 26–27, 45, 50–51, 55–50, 60–62, 64–68, 73–75, 79–83, 85–87, 90–92, 94–95, 99, 100–101, 103, 107–111, 139, 142, 144–145, communication participants 42, 51, consciousness of speaking and hearing 27, experience-/s 61, 88, goals of 95, human

(being) 24, 32, 46, 57–59, 68, 70, 93, interacting 42, 95, 101, language bearers 17, linguistic structure of 28, members 34, monolingual, bilingual or multilingual 76, 79, needs, 52, 88, organisms 32, participants 26, persons 34, properties of 13, 28, 82, 112, realizations 112, speaker/hearer 56, 120, speakers 27, 120, speaking 34, 77, 118, 130, speaking and understanding 15, subjective universe of an 25, tasks of 16, 113, user 24, 42
Ingold, Tim (b. 1948) 81, 88
instrumentalism, utilitarian 30, functionalist 30, 106
integrationist, conceptions 5, 24, critique of orthodox linguistics 24
internal, arrangements 66, compartmentalization 128, 140, conditionings 73, dimension 9, organization of language 52, and external linguistics 120, stability of a living system 30, stigmatization 118, structure-/s 84, 116, world model 24
interpersonal, and intersubjective linkages 7, 105, 108, 111, communication 16, 21, 26, 46–47, 57, 65, 69, 76–77, 88, 92, 96–97, 100, 112, 129, function 23, 45, linkages 13, 51, 56, 68, linkage systems 55, 79, relationships 7, 12, 66, 101, 122, 137, tasks 107, understanding 23, 88
intersubjective, functions 6, 45, world 6, 12, 55, 59–60, 69–70, 113, knowledge 99, 109, linkages 7, 51, 105, 108, 111, relationships 55, 109
intersubjectively, comprehensible 56, shared social life 60
intersubjectivity 59, 70
Ivić, Milka (b. 1924) 20
Jakobson, Roman (Osipovič) (1896–1982) 6, 42–44

Janse, Antheun 124–125, 132, 134, 137–138, 140
Jansma, Lammert Gosse 129, 141
Jensma, Goffe (Th.) (b. 1956) 132–133
Jesus of Nazareth, titled in Greek as Christ (*Christós*), which corresponds to the Hebrew *Messiah* (b. 4–2? B. C., crucified A. D. 29–36?) 127
Jonkman, Reitze Jehannes 130–131, 137, 139, 142
Joseph, John E. (b. 1956) 20
Kant, Immanuel (1724–1804) 103
Kelly, George Alexander (1905–1967) 57, 69
Kielar, Maciej (b. 1978) 137
Knapp, Mark L(ane) (b. 1938) 98
Koerner, E(rnst) F(riderik) K(onrad) (b. 1939) 44
Korżyk, Krzysztof 24, 56
Kress, G(unter) R. 45, 47
Kretschmer, Paul (1866–1956) 18
Kull, Kalevi (b. 1952) 80, 88
Kuryłowicz, Jerzy (1895–1978) 22
Labov, William (b. 1927) 20
Lamb, Sydney M(acDonald) (b. 1929) 23
Landsberg, Piotr 56
language, bearer-/s 17–18, 24, 76, 77, 109, 115–116, 123, change-/s 17–19, conception of 56, contacts 20, 109, entities 21, 25, evolution 17, norm-/s 20, 78, planning 24, 78, 116, policy 78, 116, 119, user-/s 5, 22, 39, 51, 77, 109, 115, 121
Leech, Geoffrey (Neil) (b. 1936) 69–70, 95
linguistic, anthropology 24, 26, 70, communication 25, 58, 64, 69, 79, 87, 93, 99, 107, communit-y/ies 13, 26, 51, 56, 58–59, 68, 79, 86–87, 90–91, 95, 107–108, 110–113, 117, 137, geography 20, linkages 5, 6, 9, 65–66, 69, 79, 99, 111–113, 139, pragmatics 6, 24, 91, 95–96, 100, properties 6, 12, 15–16, 28, 55–56,

63–67, 74–75, 78–80, 82, 85, 87, 89, 92–94, 107–112, 138, 145, sign-/s 22, 40–41, 45, 112, social and cultural context 24
linguistics, anthropological 24–25, 70 applied 27, 56, 109, cognitive 24–25, 70, comparative 11, 17–18, human-centered 12, 28, 80, 83, 85, 101, 107–109, 111–112, 137, human linguistics 6–7, 12, 28, 56, 64–66, 91–95, educational 24, critical 24, text 24, poststructuralist 24
linkage-/s, aggregations 6, 73, communicative 12, 13, 70, 82–83, 109–112, composite 95, coupled 66, 138–139, ecological 7, 137, Frisian 138–139, 141, 144, Frisian-speaking 12, 144, focused 95, global 6, 73, human 12–13, interacting 85, interpersonal 7, 13, 51, 55–56, 68, 79, interpersonal and intersubjective 7, 105, 108, 111, intersubjective 7, 51, 105, 108, 111, linguistic 5–6, 9, 65–66, 69, 79, 99, 111–113, 139, linguistic-communicational 13, long-lasting 28, 85, 90, 93, 137–138, 141–142, local 87, local and global 6, 73, semiotic communicational 110, social 44, 50, 85, 110, 138, task-oriented 111, temporary 112, temporary and long-lasting 28, 90, 93, 142, rumor-kind 140
Lizis, Elżbieta (zob. Wąsik, Elżbieta) 11, 12
local, and global 6, 73, 89, 108, communication 141, communities 87, dialect 119, ecosystems 90, environments 87, 110, individuals 86, flora and fauna 90, interest groups 86, languages 86, linkage-/s 6, 74, 87, stability 90, varieties 117, vernacular-/s 76, 78, 117
Lockwood, David (b. 1940) 23

logical, domain-/s 6, 66–67, 69, 88, 91–93, 101, reasoning 69, 94, relationship 46, semantic structure 98
logocentrism 16
Loos, Adolf (1879–1933) 31
Lorenzer, Alfred (1922–2002) 31
Luckmann, Thomas (b. 1927) 58–60
MacDonald, Graham 83
Magnus Forteman (c. 809), 126, 139–140
Makkai, Adam 81
Malinowski, Bronisław (Kasper) (1884–1942) 24, 33, 44, 46–47
Mathiot, Madeleine 52
Maturana, Humberto R. (b. 1928) 25, 70, 88
McLuhan, Marshall (1911–1980) 87
Mead, George Herbert (1863–1931) 58
meaning-/s, disproportionate 52, and understanding 88, as function in context 46, ascription of 57, 66, at all 28, bearers 13, 51, 55, 58, 60–62, 67, 69, 74, 101, 107–108, 110–113, carry 59, cognizers 51, 55, creation 61, "enactive" concept of 70, information and function 48, interpreters 51, 55, intersubjective world of 6, 12, 55, 69, 113, knowers 51, 55, maker of 61, negotiate the 71, -negotiating activities 61, objective 25, 71, of linguistic units 71, of phenomena 61, of speech acts 46–47, of symptoms 136, of the context 36, of verbal messages 55, of words 20, 48, or force of utterances 69, personal 61, producers 51, 55, referential 57–58, search for the 96, searches for 61, semantic 48, 98, shared 62, 113, sign- and- 101, social 25, 70, subjective 59–60, 145, symbolic 133, to perform 36–37, understanding of 25, 71
Meer, Geart van der (b. 1944) 10
member-/s, indigenous 121, of a collectivity 68, of (a/this) (homogenous/

communicative/discursive/ethnic and national, linguistic/local linguistic/a particular/certain linguistic/speech) communit-y/ies 23, 26, 26, 86, 45, 49, 76, 135, 58, 107, 60, of commutes 108, of (a) societ-y/ies 15, 24, 34, 59, 68, of (a/the/this) (societal/ethnic and national/small/ other/interacting/different) group-/s 26, 27, 45, 65, 84, 86, 93, 94, 118, 124, 127, of group linkages 108, of linguistic linkages 111, of the *Fryske Akademy* 141, of (Frisian-speaking and non-Frisian-speaking, a monolingual, such) association-/s 143, particular 22, 59

mental, activit-y/ies 32–33, 51, and social device 23, cognition 51, contents 32, construction 61, equivalents 51, fact 69, 93, forms 50, functions 33, interior 20, intersubjective knowledge 109, life 43, model-/s 61, processing and interpreting 62, property 27, reception 108, reflections 25, 70, relationship-/s 55, 88, 92, schemata 61, spaces 107, states 36, 102, traits 15, world 61

mentalism 20, 102

mentalist, constructivism 25, view 23,

Meringer, Rudolf (1859–1931) 20

Merton, Robert K(ing) (1910–2003) 33–34

message-/s, (as a constituent) 42–44, 46, 96, domestication of their 87, exchange of 26, global 98, interpretation of 44, production 112, receivers 45, rumor-kind 139, sending/send and receiving/receive 26, 51, 55, 108, textual 78, verbal 13, 55, 109

Milewski, Tadeusz (1906–1966) 41

Misiak, Małgorzata 11

Moffett, James 61

Moles, Abraham André (1920–1992) 31

Morciniec, Norbert (b. 1932) 10

Mühlhäusler, Peter 81

Mukařovský, Jan (1891–1975) 44

Nagel, Ernst (1901–1985) 34–35

national, anthem 135, consciousness 7, 128, 134, (and international) character 55, 133, communities 26, hero 139, identity 118, 120, 128, 137, language 49, minority 119, means of 120, pride 120, scale 134, symbol 51, 135

natural, and cultural settings 73, and/or cultural object 50, and psychical domains 102, and socio-cultural domains 75, barriers 89, biological order 89, body parts 35, changes 32, (and cultural) ecosystems 83, 86, 89, 110, end 32, (and cultural/social/socio-cultural) environment-/s 13, 19, 33, 55, 68, 70, 88, 107–108, functions 49, language-/s 47, 52, 73, 76, 115, level 89, or cultural goods 87, 91, phenomenon 50, selection 32, sciences 18–19

naturalism 19

naturalist, and materialist approaches 19, evolutionism 32, heritage 75

Nęcki, Zbigniew (b. 1947) 57

Neurath, Otto (1882–1945) 102

Newton, (Sir) Isaac (1642–1727) 103

Nöth, Winfried (b. 1944) 75, 81

object, of linguistic/material/scientific studies 5–6, 16–17, 21, 24, 27, 50, 56, 63, 79, 109, 120, of investigation 15, 25, 93, 106, of linguistics proper 79, of reference 57

objective, facts 59, functional consequences 34, meaning 25, 71, knowledge 62, 69, properties 102, world 59, 62

objectively, available components 67, 145, observable 36, 102, false categorization 84

Oertel, Hanns (1868–1952) 18

organicism 30, 34, 106

Names and terms from the main text 173

organicist-/s 30, functionalism 30, 106, notion 33
organism-/s, as a whole 32, 36, 49, biological 15, 19, 25, 35, 70, goals of an 32, 35, human being as an 59, lived histories of 88, living 19, 36, needs of 5, 33, -oriented studies 25, parts of 105, phenotypic features of an 88–89, potentials of 32, -related functionalism 106, state of 102
Osthoff, Hermann (1847–1909) 18
Park, Robert Ezra (1864–1944) 75
Parsons, Talcott (1902–1979) 33–34
participant-/s, (linkage constituent) 65–66, communicating 65, communication 12–13, 25, 42, 51, 55–56, 61–63, 66–67, 69, 79, 82, 87, 93–94, 96, 99, 101, 107–108, 110–111, 145, covered or overlapping 95 in social reality/actions/roles 12, 82, in/of (interpersonal/global/social/group) communication 26, 28, 44, 64, 66–67, 71, 77, 80, 82, 86, 92–93, 99, 110, 141, in/of (the so-called, communicative) linkage-/s 65, 94, in/of the West Frisian /Frisian-speaking/the Frisian linkage-/s, 138, 144, of communicational events 111, of group/social interactions 64, 110–111, of social interactions 110
pattern-/s, cultural 34, 82, discursive 111, of (interpreted) discourse-/s 46, 74, 80, genres or 79, of (verbal/animal) behavior 88, 112, 122, of concatenated forms 23, of study 122, of their interpreting and understanding 111, of verbal conduct 109, or practices 34
Pettit, Philip (b. 1945) 83
physical, activity 33, and chemical processes 36, and logical 6, 91, and psychical 55, 57, 107, bodies 102, constituents 83, contact 42, discontinuity 84, domain 16, 66, 69, 92–93, 100–101, 145, nature 51, objects 28, 67, 69, 94, 101, 145, parts 88, 92, proximity 84, reality 57, sound 69, 92–93, 101, surrounding 65, tests 65, working conditions 31
physicalism 102
physics 67, 93, 99, 102–103, 108, 145
Piebenga, Jan Tjittes (1910–1965)) 126, 137, 140
Pietraszko, Stanisław (b. 1928) 85
Pliny, the Elder (Gaius Plinius Secundus, A.D. 23–79) 124
Pobojewska, Aldona (Teresa) (b. 1949) 68
Podsiad, Antoni (b. 1932) 102
Popper, (Sir) Karl (Rajmund) (1902–1994) 52
poststructuralism 16
poststructuralist, attitude 113, conceptions 25, integrationism 24, linguistics 24
practice-/s, changeable 74, 80, 111, communicational 109, 112, group 84, investigative 18, 78, discursive 25, 79, 87, 111, linguistic 88, of message production 112, of verbal communication/behavior 96, 107
principle of abstractive relevance 5, 6, 22, 30–31, 40
prop-/s 28, 65–66, 79, 95, 138–139, 144
propert-y/ies, social 21, common 49, linguistic 6, 12, 15–16, 28, 55–56, 63–65, 67, 74–75, 79–80, 82, 85, 87, 89, 92–94, 107–112, 138, 145, linguistic-communicational 15, 67, 145, of man/people 12, 22, 55, 67, 80, 85, 89, 92, 105, 107, 109, 112, 145, of (a) communicating individual-/s 6, 12, 15, 55, 64, 67, 74–75, 79, 82, 87, 91–92, 101, 103, 110–111, 144, of task-realizing community 5, 29, personal 21, 62, subjective 16

psychological, approaches 81, design 23, as well as biological factors 59, frameworks 25, functions 33, property 6, 45, purposes 46, reality of language 46

psychology 29, 30, 32–33, human or animal 36, functional 33, functionalism in 32, functional studies in biology and 35, of personality 34

purpose-/s 16, 30–31, 34, 36–39, 44, 51–53, 84, -and-need-oriented rationalism 31, a tool for communicating 106, communicational 71, conscious and explicit 34, for and in itself 44, in nature and culture 35, in-view 45, of a polite exchange 97, of an acting subject 30, of communicating minds 46, of convincing someone 63, of everyday life 31, of interpersonal/mutual understanding 23, 76, 113, of language speakers 39, of S, 38, -oriented 38, 40, 45, 106, realization of 51

Radbod, also known as Redbad (680–719) 126–127, 139–140

Radcliffe-Brown, Alfred Reginald (1881–1955) 33

recentivism 68, 70, 113

reference, domain-/s of (cognitive) 70, 88, 90, 92, 99, 101, common 107, objects of 57, 71, 107, of speech products 41, to the knowledge 61

referential 64, 78, and interpretational activities 109, function-/s 43, knowledge 88, linguistics 66, value 56, 67, 69, 94, 98, 101, 109, 145

referentialism 22

referents 42

Riegler, Alexander 61–62

Robertson, Roland 87

Robins, R(obert) H(enry) (1921–2000) 16, 18, 24

Roblyer, Margaret D. 61

Salverda, Jan Cornelis Pieter (1783–1836) 133

Saussure, Ferdinand (-Mongin) de (1857–1913) 21, 23

Saville-Troike, Muriel (b. 1936) 25

Saxo 125

Schaff, Adam (1913–2006) 102

Schlegel, Friedrich (Karl Wilhelm) (1772–1829) 16

Schleicher, August (1821–1868) 17–19

Schlick, Moritz (1882–1936) 102

Schmidt, Johannes (1843–1901) 18

Schurer, Fedde (1898–1968) 133

Searle, John R(oger) (b. 1932) 89, 95

Sebeok, Thomas A(lbert) (1920–2001) 43

Seebold, Elmar 124

Siek-Piskozub, Teresa (b. 1949) 57, 69

Simons, Menno (1492–1559) 128, 139

Smoczyński, Wojciech (Sławomir) (b. 1945) 22

social, actions 59, 82, activity 51, actor 26, attitudes 136, behavior 58, being 58, changes 132, character 21, 41, 62, 112, classes 77, collectivities 56, communication 21, 57, 82, 92, 107, communities 85, conditionings 57, 113, construct 59, construction of reality 6, 57–58, constructivism 57, 61, 63, 68, 113, constructivists 59, 61, 69, and cultural context 24, 96, 101, conventions 112, device 23, domains 6, 48, environment 33, 57, 63, 107, existence form of language 19, factors 58, function-/s 33–34, 48–49, and cultural forces 76, gatherings 142, group-/s, 6, 34, 74, 82, 111, 142, grouping-/s 51, 74, 82–83, 111, hierarchy 117, identity 84, interactions 59–60, 110–111, institution 51, ladder 117, life 60, 107, 109, 119, 141, linkages 44, 50, 85, 110, 138, meanings 25, 70, movements 87, and culture-creative nature 31, needs 34, network 110, norm-/s 42, 44, ranking 143, pressures 26, process 62, propert-y/ies

21, 142, reality 12, 58, relations 26, 90, requirement-/s 30, 52, 69, 88, role-/s 28, 82, sanctions 22, sciences 28, 30, 106, situation 58, status 118, stratification 82, 108, 115, 122, stratum 117, studies 73, subjects 34, system-/s 33–34, norms and values 44, world 87
society 5, 7, 15, 18, 22, 24, 33–34, 36, 49, 58–60, 68, 74, 79–80, 83, 85, 128–130, 133, 136, 140–141
sociological, and demographic features 131, approaches to the existence forms of language 23, aspects of linguistic pragmatics 6, 91, constructivism 58, data 142, discourse 123, and psychological frameworks of ethnomethodology 25, and historical investigations 137, literature on Frisian identity 132, notion of autonomy 85, polls 131, perspective 121, pragmatics 12, 81, questionnaires 130, studies 75, 142, understanding of knowledge 68, usage 83
Sokołowska, Olga 98, 99
sound-/s, -chains 82, linear sequences of 23, relevant 65, 94, speech 17, 87, 94, waves 7, 15, 28, 51, 55, 60, 65–69, 92–93, 95, 101, 108, 145
source-/s, agent 88, 92, (authors, senders) 26, or destination 36, location 88, 92, value 88
speaker-/s 18, 20, 23, 26, 42, 27, 56, 63–64, 69, 74–75, 92, 94–97, 101, 109, chain of 19, of Frisian 117–118, 120, 142, non-Frisian language 122, Frisian 130, 142–143, Frisian and non-Frisian 141, 143, individual 27, 56, 120, native 119–120, 131, unstable 20
speech, act-/s 25–26, 46–47, 52, 69, 93, 95, 100, 113, community 24, genres 25, 52
Stampe, Dennis W. 99
Stewart, William A. 6, 48–49

structuralism 16, 21–22, 32, functionalist 30
structuralist, ideas 23, view of language 22
structure, of language 20, 24
subjective, aim-in-view 34, 45, cognition 15–16, dispositions 34, (experience) 62, inferences 67, 145, meanings 59–60, 145, mental states 36, 102, needs 5, 25, 33, 71, organization of their experiential domain 62, process 60, property 16, universe-/s 25, 32, 68, 71
Sztompka, Piotr (b. 1944) 58, 87
target-/s, agent-/s 88, 92, location 88, 92, value 88
Tarski (Teitelbaum) Alfred (1901–1983) 44
task-/s (along with: capacity, duty, function, intention, aim, goal, value) 16, 26, 28, 30, 36, 39, 50–51, 66–67, 94, 96, 100, 107, 120–121, 141, 145, being performed 13, common 124, 139, communicational 28, 52, 55, 64, 66, 69, 76, 82, 89, 92, 94, 108, 113, 138, concept of 138, fulfillment of 39, hierarchies of 66, 93, hierarchy of interpersonal 107, main/principal 138–139, 141, of communication participants 69, 94, of groups 139, of individuals 113, of people 28, 51, 89, of the participants 82, 100, -oriented 108, 112, 137, -realizing communities 5, 29, shared 75, signs of 13, term 93, the search for 145
teleological, (function) 41, (purpose-oriented) 38, (purpose or goal-oriented) 106, explanation 32, or etiological explanations 5, 32, sense 34, statements 35, system 33
teleology, etiology and 35, 105, instrumental 37
Thomson, Evan 25, 70

transactional, (vs. interactional) 26, analysis 26, behavior 58, unifying, etc. 52
Uexküll, Jakob (Johannes) von (1864–1944) 68
uniformitarian, evolution 19, uniformitarian theory 18
uniformitarianism 18
value-/s 25, 27, 34, 36–37, 44, 67, 71, 84, 88, 113, 125, communicative 63, 107, 109, of language facts 22, of meaning bearers 51, 67, of social conditionings 57, referential 56, 67, 69, 94, 98, 101, 109, 145, relational 21–22, 50, pragmatic 97, 99, source 88, symbolic 127, target 88
Molen, Sytse J. van der 132
Plank, Pieter H. van der 130–131
van Maerlant, Jacob (or Merlant, ca. 1225–ca. 1300) 140
Varela (García), Francisco J(avier), 1946–2002) 25, 70, 88
Vries, Oebele (b. 1947) 10
Wandt, Karl-Heinz (b. 1952) 79
Wąsik, Elżbieta (Magdalena) (b. 1961) 11–12, 28, 48, 57, 74–75, 76, 77, 82–83, 115, 123, 137
Wąsik, Zdzisław (b. 1947) 10, 12–13, 18, 20, 22, 28, 50, 56, 61, 64, 70, 76, 85, 89, 115, 137
Wawrzyniak, Zdzisław (Kazimierz) (b. 1944) 50
Weber, Max (1864–1920) 59
Wenker, Georg (1852–1911) 19
Wierzbicka, Anna (b. 1938) 80
Wilfryd, the archbishop of York from England 140
Wirrer, Jan (b. 1944) 74, 78
Wittgenstein, Ludwig (Johan) (1889–1951) 24
Woodfield, Andrew 35, 52
Worp van Thabor 126
Wright, Larry (b. 1937) 35, 53

Yngve, Victor H(use) (b. 1920) 6–7, 12, 26–28, 56–57, 63–67, 75, 79, 92, 93–96, 109, 137–138, 145
Zabrocki, Ludwik (1907–1977) 86
Zawadowski, Leon (b. 1914) 10, 41–42, 76, 115

Philologica Wratislaviensia
From Grammar to Discourse

Edited by Zdzisław Wąsik

The subject matter of this series is intended to cover a wide range of interdisciplinary research works on the *texts* or *text-processing* activities of humans embedded as communication participants into their social roles and culture. Within the scope of particular topics, the readers may find academic treaties pertaining not only to the structure and content of meaning-bearers materialized in the verbal behavior of people but also to their functioning in the domain of art and education. Respective contributions in the form of books and articles will be made by specialists of theoretical and applied linguistics, as well as the history of literature and intercultural communication engaged in the process of second language teaching.

Vol. 1   Zdzisław Wąsik / Tomasz Komendziński (eds.): Metaphor and Cognition. 2008.

Vol. 2   Elżbieta Wąsik: Coping with an Idea of Ecological Grammar. 2010.

www.peterlang.de

Barbara Lewandowska-Tomaszczyk / Tomasz Płudowski / Dolores Valencia Tanno (eds.)

The Media and International Communication

Frankfurt am Main, Berlin, Bern, Bruxelles, New York, Oxford, Wien, 2007.
442 pp., num. fig. and tab.
Studies in Language. Edited by Barbara Lewandowska-Tomaszczyk. Vol. 15
ISBN 978-3-631-56707-4 · pb. € 68.20*

The twenty-first century is witness to complex social, political, and cultural phenomena transforming the world in which we live. There are numerous aspects to this global process; most of them, however, are related one way or another to the media of communication which foster and accelerate it. The chapters in this book approach media and international/intercultural communication from various global perspectives. The authors provide insight into the impact of media on different contexts, cultures and nations. One theme that weaves its way throughout this collection of essays is an intercultural one, broadly defined as the contact point between two cultures that changes both to some degree. Scholars from different places in the world try to understand, explain and/or argue from a variety of traditions, perspectives and values. They examine the contact point between culture and identity, media and culture, art and media, technology and translation, theater and culture, etc., in order to better understand how and to what degree changes occur.

Contents: European identity · International communication · Intercultural communication · Polish media · British media · American media and journalism · Media policy · Media language · Information and communication technology · E-discourse · Text messages · Online pornography · Globalization · Localization · Globalization · Public relations ethics · Television genres · Film adaptations · Translation · Minority languages

Frankfurt am Main · Berlin · Bern · Bruxelles · New York · Oxford · Wien
Distribution: Verlag Peter Lang AG
Moosstr. 1, CH-2542 Pieterlen
Telefax 00 41 (0) 32 / 376 17 27

*The €-price includes German tax rate
Prices are subject to change without notice
Homepage http://www.peterlang.de